制造过程质量异常诊断的智能方法研究

程志强◎著

·北京·

内 容 提 要

产品的质量水平高低是影响一个国家经济发展和国际市场竞争能力的重要因素。产品质量、制造过程质量、服务质量问题近年来日益得到我国政府和广大公众的关注和重视。

本书对制造过程质量异常诊断的智能方法进行了研究，主要内容涵盖基于 PNN 的制造过程质量诊断、基于 LS-SVM 的小样本过程质量诊断、基于 Cuscore 统计量的过程质量智能诊断、多元过程质量智能诊断与异常变量识别等。

本书结构合理，条理清晰，内容丰富新颖，可供从事纳米材料研究的相关人员参考使用。

图书在版编目（CIP）数据

制造过程质量异常诊断的智能方法研究 / 程志强著
. -- 北京：中国水利水电出版社，2017.12
　　ISBN 978-7-5170-6104-5

Ⅰ. ①制… Ⅱ. ①程… Ⅲ. ①智能技术－应用－制造过程－质量控制－研究 Ⅳ. ①TB4

中国版本图书馆CIP数据核字(2017)第304527号

书　　名	制造过程质量异常诊断的智能方法研究 ZHIZAO GUOCHENG ZHILIANG YICHANG ZHENDUAN DE ZHINENG FANGFA YANJIU
作　　者	程志强　著
出版发行	中国水利水电出版社 （北京市海淀区玉渊潭南路1号D座 100038） 网址：www.waterpub.com.cn E-mail：sales@waterpub.com.cn 电话：(010)68367658(营销中心)
经　　售	北京科水图书销售中心（零售） 电话：(010)88383994、63202643、68545874 全国各地新华书店和相关出版物销售网点
排　　版	北京亚吉飞数码科技有限公司
印　　刷	北京一鑫印务有限责任公司
规　　格	170mm×240mm　16开本　19印张　246千字
版　　次	2018年5月第1版　2018年5月第1次印刷
印　　数	0001—2000 册
定　　价	91.00元

凡购买我社图书，如有缺页、倒页、脱页的，本社营销中心负责调换

版权所有·侵权必究

前　　言

　　产品的质量水平高低是影响到一个国家经济发展和国际市场竞争能力的重要因素。产品质量、制造过程质量、服务质量问题近年来日益得到我国政府和广大公众的关注和重视。目前,很多国家都将提高产品质量和制造过程的质量水平作为国际竞争的长远的关键战略。

　　产品的质量水平取决于产品制造和形成过程中的各个环节的质量水平。对产品的制造和形成过程实施持续的改进策略,是提高产品质量水平的有效途径。对产品制造过程的各个关键质量特性和状态进行监控和异常诊断、发现过程的质量异常现象,是实施过程持续质量改进的起点。质量诊断可以为过程质量的持续改进指明方向。通过质量诊断发现过程异常并查找过程异常的根本原因,从而采取相应有效的纠正措施,可以使过程恢复并保持在稳定受控状态。随着制造过程的现代化、智能化和复杂程度的提高,相应的,在对制造过程进行有效控制和质量水平诊断方面也提出了更高的要求,面临更多的问题和技术手段等方面的挑战。就目前来看,单纯使用传统的基于统计学的质量诊断技术并不能很好地满足这些新的要求。因此,在进行质量异常诊断的过程中引进并综合使用包括计算机、人工智能等其他技术领域的最新技术研究成果,代表了质量诊断技术的重要发展方向,也是实现智能制造和制造产业水平升级的重要手段。本专著针对这一问题展开了重点研究,将计算机人工智能技术应用于解决制造过程的质量诊断问题,以便诊断和发现制造过程中的异常。本书涉及如下的主要研究内容:

　　(1)基于概率神经网络的控制图模式识别。传统单变量控制图是诊断过程异常的重要工具,但对于过程中出现的控制图模式现象却无法加以正确判断。本书在对概率神经网络的结构特点进行研究的基础上对其参数进行了设计,提出使用概率神经网络对控制图的各个模式进行模式识别,并通过仿真实验对使用概率神经网络进行控制图模式识别的性能进行了评估,解决了使用其他类型神经网络识别控制图模式时存在的神经网络结构设计困难且识别率低的问题。

(2)基于最小二乘支持向量机的控制图模式识别。传统的SPC(Statistical Process Control)过程异常诊断方法建立在统计学大数定理基础之上,其过程诊断结论只在大样本条件下才有效,而使用神经网络技术诊断过程异常时也需要使用大量的训练样本。当可以获取的样本数量有限或者获取大量样本的成本过高时,这两种方法并不适用。为此,本书提出使用最小二乘支持向量机技术进行的控制图模式识别,并对其性能进行评估。同时,为提高模式识别的性能,提出使用粒子群算法和遗传算法优化最小二乘支持向量机的参数,实现了在有限样本条件下控制图的有效识别。

(3)Cuscore统计量对过程中预期异常信号的诊断。根据制造过程积累的先验知识,某些过程异常信号具有可预期的特征。使用Cuscore统计量诊断过程中的预期异常信号,可以有效地利用以往制造过程中积累下来的关于过程异常的先验知识,提高诊断的效率。本书研究并评估了Cuscore统计量用于诊断非线性二次预期信号时的性能;提出了使用移动窗口技术和最小二乘支持向量机模式识别技术进行变点检测的方法,解决了标准Cuscore技术中存在的失配问题,提高了Cuscore统计量对于预期异常信号的检测能力。

(4)多元过程质量诊断及异常变量识别。实际的制造过程多数属于多元过程且各个变量之间常常存在某种相关性。目前的多元SPC技术只能诊断制造过程的整体质量状态,并不能对异常变量进行有效分离和明确定位。结合传统多元SPC技术,本书分别构建了多元过程均值矢量和协方差矩阵的智能诊断模型,将均值矢量和协方差矩阵的异常变量辨识问题转化为模式识别问题来加以解决;设计了最小二乘支持向量机模式识别器;对提出的两个异常诊断模式进行了性能评估;在诊断多元过程总体质量状态的基础上,实现了对于异常变量的辨识;通过对二元过程实例的分析,具体说明了模型的应用方法。

客户的需求在当今社会日益呈现出多样化的特点。由于客户的需求不同,对生产的模式和要求也是不同的。针对制造过程的不同特点和不同的过程质量诊断要求,本书在研究以往过程诊断技术的同时,将人工智能技术应用于过程质量诊断,弥补了以往过程质量诊断技术中存在的不足。本书的研究内容所解决的关键问题主要包括:

(1)使用最小二乘支持向量机进行控制图模式识别,实现了在小样本条件下制造过程的异常诊断。休哈特控制图在大样本条件下才具备良好的过程异常诊断性能,在小批量生产模式下由于生产样本难以获取,给基于传统 SPC 技术的质量诊断工作带来困难。本书提出使用基于统计学习理论的 LS-SVM 技术来诊断控制图异常模式,实现了在有限样本条件下的过程异常诊断。

(2)提出使用移动窗口法和 LS-SVM 模式识别技术进行过程变点智能检测的模型及其实现方法,解决了标准 Cuscore 控制图理论中存在的失配问题。标准的 Cuscore 控制图技术中的"失配"问题会降低 Cuscore 图诊断过程预期异常信号的效率,本书使用移动窗口法将过程的变点检测问题转化为异常模式的识别问题,从而可以为 Cuscore 控制图提供触发信号的合理估计值,有效地解决了标准的 Cuscore 控制图中存在的失配问题,进一步提高了诊断效率。

(3)提出多元过程均值矢量和协方差矩阵异常诊断的智能模型与实现方法,在诊断多元过程总体质量状态的同时,实现了对于多元过程异常变量的识别。传统的基于 MSPC 技术的多元过程质量诊断只能够诊断过程的总体状态,不能具体识别和分离异常变量。本书以模式识别技术为手段,通过研究并定义均值矢量和协方差矩阵的模式,将统计学多元过程均值矢量和协方差矩阵的异常诊断问题转化为模式识别问题,解决了多元过程异常变量的识别与定位问题,弥补了传统 MSPC 技术诊断过程异常的缺陷。本书的研究通过这些新的智能技术手段和方法的合理应用,增强了过程质量诊断的效力,为质量诊断提供了新的研究思路和实现技术,对于提高制造过程的质量诊断水平有着现实的指导意义。

本书是作者博士研究工作的一个总结,是以作者的博士论文和发表的相关 10 余篇 SCI、EI、CSSCI 等期刊论文为基础,并结合相关理论和实践研究资料而著述成书的。本专著的出版得到了华北水利水电大学高层次人才科研启动项目(编号:201102043)的资助,在此表示诚挚感谢。

限于作者的研究水平、时间和精力,所得到的研究结论仅属于阶段性成果,疏漏和不足之处在所难免,诚恳希望同行专家、研究人员和其他读者不吝给予批评和指正。

程志强
2017 年 10 月

目 录

目录

第 1 章 绪论 …………………………………………… 1
    1.1 质量诊断技术的研究意义 …………………………… 1
    1.2 质量诊断 ……………………………………………… 3
    1.3 质量诊断技术的发展及国内外研究现状 …………… 6
    1.4 本书的研究内容、结构和研究方法 ………………… 18

第 2 章 基于 PNN 的制造过程质量诊断 ………………… 21
    2.1 过程异常与控制图的使用 …………………………… 21
    2.2 控制图模式识别问题 ………………………………… 27
    2.3 神经网络与控制图模式识别 ………………………… 32
    2.4 基于 PNN 控制图模式识别的过程异常诊断 ……… 38
    2.5 本章小结 ……………………………………………… 54

第 3 章 基于 LS-SVM 的小样本过程质量诊断 ………… 56
    3.1 制造过程的小样本质量诊断问题 …………………… 56
    3.2 有限样本条件下的统计学习理论 …………………… 59
    3.3 支持向量机理论 ……………………………………… 64
    3.4 基于 LS-SVM 控制图模式识别的过程异常诊断 … 71
    3.5 基于智能进化算法和 LS-SVM 的过程异常诊断
        技术 …………………………………………………… 77
    3.6 本章小结 ……………………………………………… 89

第4章 基于 Cuscore 统计量的过程质量智能诊断 91
4.1 Cuscore 统计量与过程异常诊断问题 91
4.2 Cuscore 统计量对于非线性预期异常信号的诊断性能 96
4.3 解决 Cuscore 图失配问题的智能变点模型 100
4.4 本章小结 118

第5章 多元过程质量智能诊断与异常变量识别 120
5.1 多元过程质量诊断问题 120
5.2 多元统计过程控制图 122
5.3 多元过程均值异常诊断与变量识别的智能诊断模型 126
5.4 多元过程散度异常诊断与变量识别的智能模型 138
5.5 本章小结 147

第6章 总结与展望 149
6.1 研究内容总结 149
6.2 展望 151

附录 以第一作者身份发表的主要学术论文 153

参考文献 277

后记 295

第1章 绪　　论

1.1　质量诊断技术的研究意义

质量是一个全社会共同关心的话题。美国著名质量管理专家朱兰有句名言:"生活处于质量的堤坝后面"(Life behind the quality dikes)。这说明产品质量问题对于社会经济生活的重大影响。从企业参与市场竞争与自身生存与发展的角度来看,企业之间的竞争要素包括企业战略、价格与成本、售后服务、品牌形象、资本与效益等,但支撑这些竞争要素的最为基础的要素就是质量。质量是企业参与市场竞争并取得长期生存的基础。人们对于质量的重视使得质量的概念和质量管理的方法不断得到发展,目前质量管理已经经过了质量检验阶段、统计质量控制阶段、全面质量管理阶段、标准化质量管理阶段、数字化质量管理阶段等五个阶段的发展[1,2,3,4]。在质量管理的发展过程中,人们不断将数理的方法应用于质量管理,其中以休哈特SPC(Statistical Process Control)控制图为代表的过程质量诊断技术的使用,大大地提高了制造企业的质量保证水平,使得人们认识到SPC质量诊断技术对于保证产品和过程质量的重要性。

20世纪90年代末以来,随着全球经济一体化以及国际贸易的快速发展,市场需求多变,企业之间的国际化竞争日趋激烈,各国企业为了提高客户满意度和取得竞争优势,纷纷根据市场的要求而改进生产模式和开发多种先进的现代化生产系统。现代工业制造过程的特点是规模巨大,制造技术及工艺复杂,随着现代化生产系统的日益复杂,人们发现在生产模式多样化、生产工艺与流程复杂化的条件下,不同特点的制造过程对于质量诊断技术的要求是不同的,影响到过程及产品质量的因素众多,单纯依靠统计学的方法并不能对复杂制造过程中的异常实施有效诊断,以传统统计学为理论基础的过程质量诊断技术在取得研究和应用成果的同时,在很多情况下并不能够满足和适应先进制造过程对于过程质量诊断提出的新的要求,人们意识到制造过程质量诊断工作中必须不断吸收和引进其他学科中先进的成果,站在系统的高度,对制造过程实施多方位立体诊断。同时,随着科技进步的快速发展,在其他学科出现了很多优秀的技术成果,这些技术成果典型地包括神经网络、支持向量机、进化优化算法、小波理论、模糊数学以及灰色理论等,它们的技术特点可以用来更好地解决制造过程质量监控与诊断问题,特别是计算机软/硬件技术、计算速度的飞跃式提高以及数据采集与存储技术的快速发展,使得这些技术也具备了在制造过程现场进行使用的条件。目前质量诊断技术已经远远突破了统计学的框架,包含了更多现代化的、先进的工程技术手段,成为一门交叉性很强的学科。本书正是在这样的市场经济背景和科技背景条件下,在传统的统计学质量诊断理论的基础上,使用计算机技术和人工智能方法,针对产品制造过程中的异常诊断问题,研究解决制造过程质量诊断问题的新途径和新方法。需要说明,我们使用人工智能的方法对制造过程进行诊断分析,不是对传统统计学过程诊断方法的否定,而是适应市场环境、制造环境以及技术环境的变化,在原有统计学方法的基础上,对原有的过程质量诊断方法做出了有益的补充和改进。

　　从理论研究的角度来看,制造过程质量诊断技术是质量工程的重要组成部分。质量工程(Quality Engineering)的概念最初由日本田口玄一[5,6](Genichi Taguchi)提出。田口博士在研究和定义质量损失函数的基础上,独创了所谓的"田口方法"。目前,质量工程的概念已经远远突破了最初田口方法的范围,其内容和方法都得到了极大的拓展。质量智能化诊断技术将计算机科学、人工智能技术等其他学科先进的成果应用于制造过程质量诊断,大大丰富了质量工程的内容,本书的研究在这方面具有重要的理论意义。

从实际应用的角度来看,对制造过程进行监控和实施制造过程质量诊断,是实现质量持续改进的有效途径之一。质量策划,质量控制和质量改进等一系列质量管理活动都需要包含质量诊断这一重要工作环节。通过对制造过程进行质量诊断,可以发现制造过程中的质量问题、查找和排除造成质量异常的原因。据工业统计,60%~70%的产品质量问题来源于制造过程,本书研究如何通过使用人工智能方法对制造过程中的异常进行监控和诊断,保证过程处于稳定受控的工作状态,从而保证最终产品的质量、安全和可靠性,提高制造过程对于异常的诊断效率。智能化过程质量诊断技术对于企业降低废品、提高质量保证水平、增加企业效益、增强企业竞争力等方面,具有重要的应用价值。

本书的研究是国家自然科学基金重点项目:"面向复杂产品的质量控制理论与方法"(70931002)和国家自然科学基金项目:"六西格玛管理的保障机制和技术研究"(70672088)的重要组成部分。

1.2 质量诊断

1.2.1 质量诊断的内容

质量诊断的概念包含了丰富的内容,就本书而言,重点研究产品制造过程质量诊断。通常,质量诊断可以看成是使用一定的管理学手段和技术,或者根据一定的标准,对企业的产品和服务、过程或质量管理工作进行考察、衡量并加以判断,以明确该产品和服务的质量是否满足规定的标准要求或者客户要求,或与该产品的形成有关的质量管理工作是否适当、有效,并进一步辨识和查找造成质量问题的根本原因,指出产品质量的改进方向,并提出相应的改进的途径与措施的全部活动。

质量诊断的内容可以大致分为三个大的方面[7]:产品质量诊断、过程质量诊断和质量管理诊断。产品质量诊断,即针对产品本身不定期或定期地对市场上出售或库存的产品进行抽样检查,确定产品质量能否满足用户的需求,掌握产品的质量数据和信息,以便采取措施加以改进。过程质量诊断就是通过对制造过程的质量进行定期检查,掌握制造过程的质量状态是否稳定,过程中是否存在异常干扰因素,过程能力是否充足,查找导致工序异常的原因以及异常问题的发生部位和发生程度。过程质量诊断过去也通常被称为工序质量诊断。质量管理诊断包括对于ISO 9000质量保证体系[8]的符合性诊断、卓越质量奖诊断、平衡计分卡方法诊断等。其中,质量保证体系的诊断就是对PDCA(策划—实施—检查—处理,Plan-Do-Check-Act)循环的诊断。对质量保证体系的有效性和符合性进行的诊断通常称之为审核(Audit),可以由组织的外部人员(如顾客或者外审员)进行,称为"外审";也可由企业内部人员进行,称为"内审"。

进行质量诊断可以使用定性分析法或定量分析法。传统的质量诊断分析方法常用的有评分法、标杆对比(Benchmarking)、平衡记分卡法(Balanced Score Card, BSC)、调查表法、Pareto图法、鱼骨图法、SPC控制图、过程能力指数分析等。

质量诊断的目的在于通过使用各种质量工程技术手段,监控并确保影响产品及其生产过程的各种因素处于受控(In Control)状态[9,10]。全球生产力和经济的发展使得生产系统日益成为一个复杂系统,其信息具有不确定性、模糊性及混沌性等特点,难以找到精确的数学模型进行描述;同时,客户需求的日益多样化使得很多生产过程具有小样本、复杂产品(如,航天器,船舶等)甚至是单件生产的特征,因此,传统的基于统计学的质量诊断理论在这种情况下已经不能满足质量管理的要求。随着科学技术的进步,各种计算机技术、人工智能技术、人工神经网络、进化优化算法、模糊理论和灰色理论、小波理论、支持向量机、混沌理论等众多的工程技术手段被不断成功地应用到质量诊断领域中,大大丰富了质量诊断的手段。目前,质量诊断理论和技术已经形成为一门综合性质的交叉学科,涉及行业及企业的各个层面,可以对影响企业发展和竞争力提高的各种质量因素进行诊断,根据质量诊断的结果查找造成质量问题的原因并采取纠正措施,从而使得企业的质量管理水平得到持续改善。

1.2.2 质量诊断的特征

产品或者制造过程的质量诊断问题,通常被称为"质量故障诊断"(Quality Fault Diagnosis)问题。对于产品或者过程的质量诊断,根据被诊断对象的不同,可以在不同的层次上进行,层次性是质量诊断的重要特征之一。质量诊断的层次通常可以分为系统级、子系统级、元部件级的故障诊断[11,12]。质量管理学中的质量诊断,诊断对象一般是"人员、机器设备、物料、制造工艺、生产环境、测量方法与设备",关注的是在系统和子系统级别上对生产过程能否满足规定的要求进行分析判断,比如,过程中是否存在异常因素,过程是否受控,过程能力指数是否满足客户要求,等等。对于部件级别和元件级别的质量故障诊断问题,多数需要根据产品所涉及的具体学科,在掌握和分析生产工艺、设备或者产品的物理和化学机理模型的基础上,借助于专业知识和专业的检测分析仪器,对故障的原因和部位加以判断并进一步制订解决方案,各个层次的质量诊断问题需要技术工程师的密切参与才能得到很好解决,单纯依靠统计学或者计算机数据分析的方法,诊断效果一般不理想。低层次的质量故障问题一定会导致高层次的质量故障问题;高层次的质量故障问题可以是由低层次的质量故障问题引起的,也可以是在不存在低层次的质量故障问题的情况下发生。

质量诊断除了具有层次性特征之外,还包括相关性、随机性、可预测性等特征。相关性是指故障之间的相互影响,以及故障及其特征之间的复杂对应关系。相关性导致了故障诊断的困难。随机性是指质量故障的出现时刻一般没有规律可循,故障信息往往具有模糊性和不确定性。可预测性是指多数的故障在其发生之前都会有一定的征兆,如果及时检测和分析这些故障特征信息,质量故障问题在很大程度上是可以预测和预防的。质量故障诊断问题的随机性、相关性特点使得造成质量故障问题的各种原因之间的关系十分复杂,传统的故障诊断方法已经不能满足现代质量故障诊断的要求,必须采用智能故障诊断的方法[13,14]。智能故障诊断技术运用人工智能技术和专家知识,不需要建立过程的精确数学模型,以人类思维对于信息的加工和认识过程为推理基础,通过对故障信息的有效获取、传递和推理分析,找到质量特性和过程参数变量之间的关系,从而对过程中的被监控对象的运行状态和故障部位进行正确判断和决策。

1.3 质量诊断技术的发展及国内外研究现状

1.3.1 质量诊断技术的发展阶段

质量诊断技术是用于控制质量特性指标和诊断过程变量异常变化的一系列理论、方法和技术的综合。在现代化生产条件下,质量诊断技术是现代工业企业保证其质量水平的一项关键性技术。质量诊断技术的发展阶段与质量管理的各个发展阶段是密切相关的,但又不完全相同。从质量监控与诊断的发展来看,主要经历了以下三个发展阶段。

1)原始质量诊断阶段

19世纪末到20世纪40年代。这一阶段的质量检验依靠的是对产品的全检或抽检,对于质量异常问题的解决多数情况下需要依赖专家经验,根据专家的直觉和经验对质量问题做出判断和改进。

2) 全面统计过程控制阶段

20世纪40年代至80年代。在这一阶段,日本和美国率先将SPC技术应用于生产,并大幅提高了制造过程的质量水平;随后,其他国家相继在其生产中使用SPC技术作为质量监控与诊断的手段。随着SPC技术的推广应用,SPC技术也获得了不断改进,并由单变量SPC发展到多变量SPC、统计过程诊断(Statistical Process Diagnosis,SPD)[15,16]和统计过程调整(Statistical Process Adjustment,SPA)[17,18]。SPD技术不仅要判断过程是否失控,还要确定与辨识异常的类型以及异常发生的位置、原因等方面的信息。SPA需要在诊断过程失控的基础上,进一步提供如何对生产过程进行调整的方案,使得生产过程重新处于稳定状态。80年代初,我国学者张公绪提出了两种质量诊断理论和系列选控图,在质量诊断理论方面做出了突出贡献。

3) 智能质量诊断阶段

20世纪80年代末至今。在这一阶段,工业过程变得越来越复杂,全面质量管理理论在全世界范围得到普遍的认同和推广应用,生产过程涉及的质量特性变量大大增多,并明显具有复杂相关性、不确定性、模糊性的特征,使用质量诊断的统计学方法处理这些诊断问题变得越来越困难。随着科技水平尤其是人工智能技术的进步,智能技术为质量诊断开辟了新的道路。智能化质量诊断技术虽然已经有近30年的发展历史,但实践证明,这方面的研究还没有形成系统化完整化的理论体系,在实践上中也没能满足现代制造业对于质量保证与诊断的要求,尚有很多问题需要研究解决。

1.3.2 质量诊断技术的类型

质量智能诊断技术涉及众多学科中的前沿成果,没有一种诊断理论可以解决全部的复杂生产过程现场中的质量异常问题,需要将各种方法综合应用于制造过程的现场才能取得良好效果。目前,在国内外质量诊断的众多研究中,涉及众多的理论和方法,按照它们的特点,可以大致分为三个类型[13,19]。

1.3.2.1 基于解析模型的方法

这类方法出现最早,特点是需要建立被诊断过程和对象精确的解析模型。实际过程诊断问题中,复杂生产过程具有多变量、时变和非线性的特征,建立过程的精确数学模型非常困难,这一方法的效果在实际中受到很大限制。这类方法中的典型方法包括参数估计法[20]、基于观测器的状态估计法、等价空间法[21]、状态空间法[22-26],它们都属于使用解析模型进行质量诊断的方法。

1.3.2.2 基于信号处理的方法

这种方法不需要建立被诊断对象的精确数学模型,主要诊断思路是认为系统的输出观测值在幅值、相位、频率及相关性上同故障之间存在着某种联系,利用信号模型,如相关函数、自回归移动平均、(快速)傅里叶变换、小波变换[27-29]等,从过程观测数据中提取诸如方差、频率、相位、幅度等信息,从而判断故障的存在和故障源的位置。

1.3.2.3 基于知识的方法

现代制造业中的生产过程是极其复杂的,已经不可能使用精确的数学模型对其进行描述。同时,计算机软硬件技术和人工智能技术的发展,为质量诊断提供了新的技术实现手段。基于知识的质量诊断方法以人工智能为主要实现技术,具有广泛的理论研究和实际应用价值。其显著优势在于,可以模拟人类思维进行推理,推理规则明确;知识可用显式的符号表达,无须被诊断对象的细节知识,适用于复杂产品和制造过程;便于计算机实现等。应当注意,基于人工智能的质量诊断方法不是对原来的基于传统统计学诊断方法的否定,而是采用新的技术手段,补充和加强了原来质量诊断方法的功效。另外,基于知识的质量诊断方法存在知识获取与描述困难、诊断系统自适应能力差、知识更新与学习能力差等缺陷。这方面的问题需要进一步研究解决。

利用人工智能进行质量异常诊断所涉及的几项热点技术包括：

1) 人工神经网络(Artificial Neural Network, ANN)诊断技术[30-38]

人工神经网络通过其所在环境的刺激作用得到学习和训练,并且不断调整神经网络的参数(权值和阈值),使得神经网络具有从外界环境中学习并提高自身性能的能力,从而具有对外界环境作出相应反应的能力。将 ANN 应用于过程异常诊断的优势在于,ANN 技术不需要被处理的观测数据的先验知识(如,分布)。在质量诊断方面,ANN 技术主要用来对过程数据的异常变化信息进行分类和模式识别。但人工神经网络的方法对于小批量生产过程的质量诊断问题无能为力,在生产中适用于大批量生产的情况;另外,神经网络的结构设计过多依靠经验,并且在使用时计算量大、耗时多。

2) 基于模糊理论的诊断技术[39-41]

质量诊断需要由被监控对象的观测数据来判断过程的状态是否正常,但过程的异常状态同正常状态之间往往没有明确的界限,存在模糊的"过渡质量状态"。对于这种情况,使用经典的二值逻辑或者多值逻辑难以处理,需要使用模糊诊断技术来解决。模糊诊断理论在判断过程的质量状态时需要借助于集合论中的隶属度函数,但隶属度函数的构造缺乏严格的理论支持,往往含有一定的经验或者主观因素,如果处理不当,将会导致比较多的诊断误判。

3) 基于统计学习理论的支持向量机技术[42-47]

统计学习理论使用该方法来实现由数据空间到特征空间的非线性变换,其核函数相对简单,使得很多在数据空间难以处理的问题在特征空间得到很好的解决。基于统计学习理论的支持向量机(Support Vector Machine, SVM)技术采用结构风险最小化原理来训练样本误差,其模型结构由支持向量决定,使得 SVM 技术可以在小样本情况下对各种控制图模式进行有效识别。SVM 技术的特点符合制造企业减少样本测试成本和提高过程监控效率的要求。根据实际问题的特点,SVM 技术常常同模糊理论以及主成分分析等其他诊断方法加以结合以用来解决分类问题[48-53]。

4)进化优化算法和群智能算法[54-63]

这类技术模仿社会性动物(蚁群、鸟群、鱼群等)的自组织行为的机理,或者从仿生学的角度对问题进行数学建模并用来解决诸如优化和聚类等实际问题。使用该项技术除了可以对质量异常诊断问题以聚类的方式直接进行求解之外,在过程异常诊断领域,智能算法更多地用来解决优化设计问题,比如,优化神经网络的结构和 SVM 的参数等。群智能算法在模式识别、流程规划、过程现场控制、决策支持以及系统辨识等领域都有成功的应用,为解决上述问题找到了新的途径。

5)其他智能技术

比如,专家质量诊断系统[64-68],使用专家系统诊断过程异常时需要使用大量的推理规则,但是随机因素的存在使得各种过程异常状态之间的差别比较模糊,影响到了专家质量诊断系统的判断精度。专家系统的发展方向是将多种其他的人工智能技术融合于其中,并使用数据采集和检测技术、网络技术、计算机技术、数据库技术的最新成果,如虚拟现实技术(Virtual Reality, VR),对过程现场的质量状态实施立体诊断。

1.3.3 质量诊断技术的国内外研究现状

制造过程质量诊断中可以使用的技术是多种多样的,涉及诸多的学科领域。我们按照质量诊断的研究内容并结合解决问题所使用的技术的特点,总结了制造过程质量诊断的国内外研究现状。目前,国内外过程质量诊断的研究主要包括以下几个方面的内容。

1.3.3.1 单变量SPC方法质量诊断

休哈特单变量 SPC 控制图[69]是过程质量诊断的传统方法,但其本身也存在许多技术缺陷,比如,该方法假定过程观测数据满足正态分布,该假设条件在许多过程(比如,流程性过程)中并不能得到满足,此时使用休哈特 SPC 图诊断过程会导致较大的误判和漏判,许多质量人员针对休哈特 SPC 控制图本身的问题以及不同性质和特点的制造过程对质量诊断的不同的要求,针对单变量 SPC 过程质量诊断问题持续展开研究,主要包括:

1)面向多品种小批量制造过程的控制图

先进制造环境下客户的需求是多样化并且不断动态变化的,小批量生产成为重要的生产模式,大批量定制的生产模式并不适用于这种市场要求。休哈特控制图建立在传统统计学基础上,只适合在大批量生产过程中使用,对于小批量生产过程并不适用。如何适应先进制造环境下多品种、小批量生产的要求,使用有限的观测数据对过程实施有效的监控与诊断,是质量诊断技术需要研究解决的问题。小批量生产中 SPC 控制图[71-74]技术的使用仍然是人们研究和关注的问题。

2)诊断过程微小异常波动的累积和控制图(CUSUM)和指数加权滑动平均控制图(EWMA)

传统的休哈特控制图对于大的过程异常波动具有很好的异常诊断能力,但对于微小的过程异常波动却不是很敏感。CUSUM 图[75,76]、EWMA 图[77,78]及其各种改进形式控制图的研究,都是为了提高 SPC 控制图对于过程微小异常变化的检测能力。

3)诊断自相关性过程观测数据异常的控制图

常规控制图技术的基本假设是从过程得到的观测值彼此独立,但许多过程实际上存在着自相关现象,如,化工过程以及数据采集频率相对较快的过程。当过程存在自相关时,需要研究新的控制图方法来提高诊断的效率,减少自相关现象给过程异常诊断带来的误判和漏判。解决自相关性过程质量诊断问题所使用的典型控制图包括改进的 EWMA 图[79]、残差控制图[80−84]等。

1.3.3.2 基于控制图模式识别的单变量过程质量智能诊断

休哈特控制图使用小概率准则来判断过程是否出现异常,比如,GB/T 4091—2001 中规定了 8 种基本失控判定准则,但许多其他的异常模式并不包括在其中[70]。对于控制图模式现象,人们研究使用人工智能模式识别技术来进行质量诊断。

在控制图模式识别的研究工作中,人工神经网络(Artificial Neural Network,ANN)作为一种有效的方法得到广泛的重视。在使用 ANN 进行控制图的模式识别与分类方面,Cheng[98]分别开发了专门用于识别均值平移(向上/向下)模式和趋势(向上/向下)模式的 BP 神经网络,Chang[99]等开发模糊神经网络识别过程均值的变化,他们使用平均链长(Average Run Length,ARL)指标来评价识别速度性能,指出开发的 BP(Back Propagation)神经网络在识别较小的和中等程度的过程变化时比传统的休哈特控制图速度更快;Perry[100]和 Guh[101−103]使用 BP 神经网络对西方电气公司定义的六类典型控制图模式加以识别,但其模式识别率有待提高,Guh 的研究同时认为学习矢量量化(Learning Vector Quatization,LVQ)网络更加适用于对于混合模式的控制图模式进行模式识别;乐清洪[104]提出使用局部有监督特征映射(Regional Eupervised Feature Mapping,RSFM)神经网络实现质量控制图的在线智能诊断分析系统的框架,该框架可以实现控制图模式识别、参数估计、专家诊断分析系统和参数调整的功能。此外,自适应共振(Adaptive Resonance Theory,ART)神经网络也被用于解决控制图的模式问题。但人工神经网络技术存在着大样本训练、经验风险最小化等问题。

支持向量机[120−123](Support Vector Machine,SVM)技术以

统计学习理论为基础,可以在有限样本条件下对控制图模式进行识别。针对实际问题,为解决支持向量机应用于包括控制图在内的多类数据分类问题,使用"模糊"(Fuzzy)方法的模糊支持向量机[48-51]被提出并使用,但其隶属度的确定方法缺乏严格的理论支持,更多地带有经验的成分。有的研究将传统的 PCA(Principal Component Analysis,PCA)技术同 SVM 技术相结合[52,53]来解决控制图模式分类问题。该技术先对工序中的原始过程数据进行 PCA 变换降低其维数、突出了主要特征信息,再根据数据特征的分布特点,有针对性地设计多分类 SVM 识别器模式识别。Suykens 提出的最小二乘支持向量机(Least Squares Support Vector Machine,LS-SVM)[124-127]是标准 SVM 的一个变种,它将标准 SVM 中的二次规划问题转化为线性方程组求解,运算速度得到提高,是目前研究较多的一类 SVM。在使用智能技术来诊断过程异常方面,SVM 技术在样本数量、运算速度等方面具有明显优势。

1.3.3.3 多元过程 SPC 方法质量诊断

传统的多元统计过程控制[162]包含了多种技术,降维技术是其中的一种。降维技术的基本思想是将多变量高维数据空间投影到相对独立的低维空间,以降低分析难度,这种方法有时被称为"基于数据处理的质量控制方法"。这其中主要包括主成分分析法、偏最小二乘法(Partial Least Squares,PLS)方法[163,164],以及独立成分分析(Independent Component Analysis,ICA)方法[165,166]。主成分分析是在力保数据信息丢失最少的情况下对高维变量空间进行降维处理。主成分分析法信息提取的实质是选择几个有代表性的主元,用来解释数据中大部分变化,从而将数据按最优形式分成两个部分:噪声部分和系统部分。主成分分析一般只能用在连续过程的监控,而多元主成分分析(Multi-way Principal Components Analysis,MPCA)可用于间歇过程的监控。偏最小二乘法主要用于建立多因变量与多自变量之间的统计关系。独立成分分析(ICA)方法的目的是将观察到的数据分解为统计独立的成分,并从混合信号里恢复出一些基本的源信号。当过程中需要监控的变量个数很多时,使用 PCA/PLS 方法来简化质量诊断问题是绝对必要的,不然,实际应用中会由于过程质量特性变量的个数过多,导致无法使用其他的质量工程技术来进行过程监控与诊断。一般认为,当制造过程中的质量特性变量多于 10个时,其他的多元统计分析方法将失去效用,此时必须使用 PCA/PLS 方法来减少质量特性变量的个数。PCA/PLS 方法是寻找关键质量特性变量的方法之一。

基于降维技术的多元统计过程控制方法在实际应用中存在一定缺陷,主要是因为降维技术在推导过程中做了一定的假设,包括:

(1)各变量都服从相同的高斯分布,但在实际过程中难以严格满足。

(2)过程处于稳态时,质量特性变量之间是序列不相关的。

(3)相关质量特性变量之间的关系是线性的,PCA 和 PLS 都是线性化的建模方法,而实际工业过程本质上都是非线性的。

(4)过程参数是恒值参数,不随时间变化,PCA 是一种静态建模的方法,不适用于动态时变过程。

此外,在传统单变量 SPC 控制图的基础上,发展起来了所谓传统多变量统计过程控制图(Multivariate Statistical Process Control Charts,MSPC Charts),其中具有代表性的是 Hotelling T^2 控制图、MCUSUM 图和 MEWMA 图[167-169]等。这些多元 SPC 质量诊断技术是基于过程数据满足多元正态分布来建立检验统计量并进行过程质量监测的。Hotelling T^2 控制图用于对多元过程的均值变量进行监控。与一元 SPC 控制图要求在方差受控前提下才能讨论均值的诊断与控制问题一样,多元 T^2 控制图也是在假定过程变量的协方差矩阵保持不变的前提下讨论均值向量的控制问题。Hotelling 构造的统计量假设过程的总体参数(均值矢量,协方差矩阵)是未知,如果过程的总体参数为已知,

Hotelling T^2 统计量就是 χ^2 统计量。Hotelling T^2 控制图的一个缺陷是,它只是使用过程的当前观测数据来检测过程异常,只适合用于检测较大的过程均值变化,而对于均值出现的较小的渐进漂移或者出现某种变化趋势时,Hotelling T^2 控制图的检测能力不足。另外,当 Hotelling T^2 控制图检测到过程失控时,它不仅不能准确诊断和定位失控变量(或失控变量的集合),而且也不能辨识是均值矢量还是协方差矩阵发生了变化。针对 Hotelling T^2 控制图对于过程均值的微小异常变化检测能力不足的问题,Croisier[170]提出了 MCUSUM 图,Lowry[171,172],Mason 和 Champ[173],Woodall[174]等于20世纪90年代提出了 MEWMA 图。与 Hotelling T^2 控制图不同,MCUSUM 图和 MEWMA 图不仅使用当前的过程观测数据的信息,而且也将过程观测数据的历史信息考虑进去,因此使得 MCUSUM 图和 MEWMA 图具有检测过程小波动与渐进漂移的功能,因此它们在现场具有更大的应用价值。MCUSUM 图和 MEWMA 图都是诊断多元过程均值向量异常的,Woodall 和 Ncube 的比较研究结果认为这两种控制图在检测过程均值的灵敏度性能方面(Average Run Length,ARL)基本是相同的,但却好于 Hotelling T^2 控制图[175]。

Hotelling T^2 控制图、MCUSUM 图和 MEWMA 图都是针对过程的均值矢量变化进行监控的。相应地,另一类 MSPC 控制图是针对过程的散度变化而进行监控的,其中,具有代表性的是样本广义方差$|S|$图、W 图和$|\Sigma|$图[168],Alt 指出$|\Sigma|$图的性能最为优越,可以作为监控多元过程方差变化的首选方法;VMax 图[176]也是设计用于监控多元过程协方差矩阵变化的,但仅仅针对二元过程,VMax 图对过程异常的检测速度(灵敏度)高于样本广义方差$|S|$图。这些多元 SPC 控制图的共同缺点是要求用来描述过程特征的多元协方差矩阵为已知,这在生产过程现场并不容易做到。

Hotelling T^2 图、MCUSUM 图、MEWMA 图等控制图都是针对过程的整体状态进行诊断的多元控制图。目前,多元 SPC 控制图质量诊断技术在实际应用中普遍存在着一定的尚未解决的问题,比如,多元控制图只是能够做到使用多元过程的总体检验统计量来判断过程失控与否,并不能对过程中出现的失控给出明确合理的解释。1980年我国张公绪教授提出选控图、全控图[177]等相关概念和过程诊断方法。全控图是对所有异因都加以控制的控制图,选控图可在多元过程中选择部分异因加以控制和诊断,但选控图与全控图之间是一一对应的,需要由全控图构造出与之相对应的选控图,因此使用选控图时需要假定或者通过大量样本数据来验证过程数据的分布状况,这在制造现场同样难以做到。按照Jackson[178]以及Hayter和Tsui[179]的观点,多元过程异常诊断技术必须解决"异常变量的定位与分离"问题。就这一点来看,多元过程的质量诊断与控制问题还远远没有得到有效解决,针对不同的多元过程状况,基于传统统计学的多元SPC技术仍然处于不断发展与改进中。同时,工业制造技术的复杂性使得人们认识到不能单纯依靠传统MSPC技术进行过程质量诊断,必须不断寻找新的方法来解决多元过程质量诊断问题。

1.3.3.4 多元过程质量智能诊断技术

如前所述,传统的多元过程SPC质量异常诊断的方法各有其缺陷。比如,使用对多元过程数据进行降维处理的方法和全控图方法时,需要有关过程分布的先验知识;而MSPC Charts的另外一个关键性缺陷在于其只能对多元过程的总体检验统计量(均值、方差)进行监控,但是当多元控制图检测到过程总体检验统计量异常时,却对出现的过程异常无法进行正确的解释,更无法准确诊断、定位与分离造成过程异常的变量(或变量组合),因此无法为过程故障的排除提供必要的信息。为此,质量研究人员将人工智能技术领域的新方法应用于多元质量诊断领域。比如,Hassen[180]将"模糊"的概念应用于多元质量控制,建立了多元模糊质量控制图;小波变换技术在信号处理方面的优良性能引起了质量人员的关注,也被应用于多元过程质量诊断中[181]。在使用计算

机技术进行多元过程诊断的研究方面,神经网络方法在这一领域取得了一定的研究成果。Noorossana[182]最先将人工神经网络应用于自相关过程异常原因的诊断,并且指出人工神经网络是进行多元过程异常原因诊断的有效方法之一。Niaki[183]使用 MLP 神经网络(Multilayer Perceptron Neural Networks)对由 Hotelling T^2 控制图产生的过程异常报警进行模式识别,实现了多元过程异常变量的诊断和定位,Chen 和 Wang[184]使用 BP 神经网络对由 χ^2 图产生的多元过程异常报警实现了类似功能。Guh[185]和 Yu[186]提出使用神经网络对二元过程均值异常进行诊断,该方法不仅可以实现对于异常变量的定位,而且可以对过程均值异常的幅值进行估计,同时,使用 ARL 进行性能评估的结果表明,该方法对过程均值异常的检测速度好于传统的多元控制图(Hotelling T^2 图、MCUSUM 图、MEWMA 图)。Low 和 Hsu[187]使用 BP 神经网络对多元过程的方差(协方差)变化进行监控,该方法同传统 $|\Sigma|$ 图诊断的结果相比较,表明神经网络方法对过程异常具有更好的检测能力,但他们在设计 BP 神经网络结构时使用了较为复杂的 DOE(Design of Experiment)方法,而且该方法同样不能实现对过程异常变量进行准确定位。

除去上面的神经网络方法,SVM 技术也被应用于多元过程的异常诊断。Ben[188]等人提出的支持向量回归控制图(SVR-based control chart),仿真实验结果表明这种控制图比 MCUSUM 图更适合于检测多元过程微小均值的变化,但该方法同样具有无法准确定位异常变量的缺陷。Cheng[189]使用 SVM 技术诊断多元过程异常,该方法通过对协方差矩阵进行模式定义,将过程异常变量的异常情况同协方差矩阵的模式相联系,实现了多元过程散度异常的诊断和异常变量定位功能,但该方法仅能对有限的几种协方差矩阵异常加以定义、识别和异常变量准确定位,对于异常变量组合导致的过程散度异常的各种复杂情况,该方法有待于定义更多的协方差异常模式,以便实现异常变量的准确识别和定位;Cheng 的研究同时指出,使用 ANN 方法和使用 SVM

方法得到的多元过程异常检测与定位的效果几乎是一样的,这两种方法都可以应用于非正态过程,但 SVM 技术所需的训练样本远小于 ANN 方法。SVM 技术具有鲜明的技术优点,使用 SVM 技术进行多元过程诊断时同样具有小样本训练、训练速度快、训练结果泛化能力强的特点,而且 SVM 技术具有可以应用于非正态过程的特点,使得 SVM 技术具有更大的应用范围和更多的应用领域。

1.4 本书的研究内容、结构和研究方法

1.4.1 研究的内容和结构

质量诊断的内容是极其丰富的,结合不同的制造过程环境,本书仅针对如何诊断产品制造过程中的异常现象这一问题展开研究,研究内容包括以下几个方面。

第 1 章:绪论。介绍了本书的研究内容、方法和意义;介绍了关于质量诊断与技术的相关内容;概述了过程质量诊断技术的国内外研究现状,探讨了这些技术中存在的问题。

第 2 章:研究基于概率神经网络的质量异常诊断技术。针对以往使用人工神经网络进行控制图异常模式识别时存在的神经网络结构设计困难,以及由此导致模式识别率不高等问题,提出和构建具有可比性的、易于设计实现的概率神经网络质量异常诊断模型,并通过计算机仿真、使用识别率、ARL 等指标对 PNN 模式识别技术的诊断性能进行评价。

第 3 章:研究在小样本条件下控制图模式识别与诊断的方法。针对小样本过程质量诊断问题,提出使用最小二乘支持向量机进行质量控制图异常模式识别与诊断的方法,并对最小二乘支持向量机的技术参数采用粒子群算法、遗传算法进行优化,提高

了过程异常诊断的性能,通过仿真实验对提出的方法进行性能评估。

第 4 章:研究 Cuscore 统计量对于非线性二次预期异常信号的诊断性能;重点研究使用移动窗口法和 LS-SVM 模式识别技术,对预期异常信号的出现时刻进行智能变点检测,以解决 Cuscore 统计量在检测预期异常信号时存在的失配(mismatch)问题,提高 Cuscore 统计量的检测性能。

第 5 章:研究对多元自相关过程进行智能异常诊断的模型。研究重点是,分别针对二元过程的均值漂移(mean shifts)和方差漂移(variance shifts)建立基于 LS-SVM 模式识别技术的智能诊断模型,实现了多元过程异常变量的定位与分离,通过计算机仿真对提出的模型和方法进行性能评估。

第 6 章:总结与展望。总结了本书研究的内容以及不足之处,并指出了需要进一步解决的问题。

本书的组织结构与内容安排如图 1-1 所示。

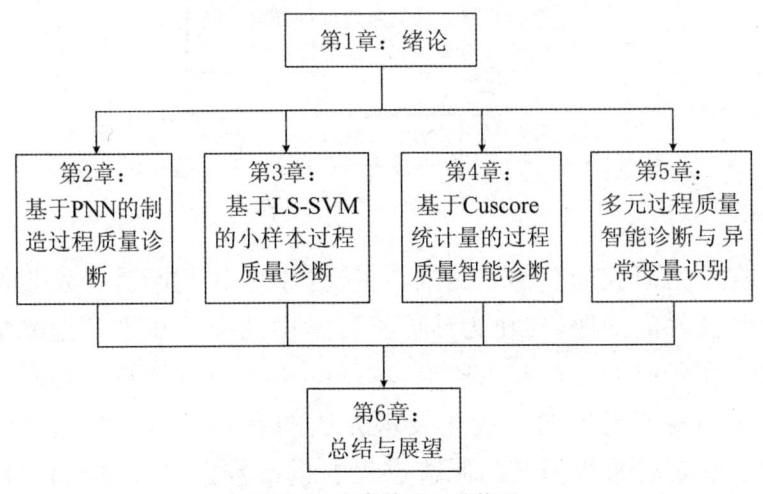

图 1-1　本书的组织结构图

1.4.2　研究的思路和方法

根据本书研究的内容、技术特点与实现方法,下面给出了本书解决制造过程质量问题,开展有关研究工作的过程、思路与方法,如图 1-2 所示。

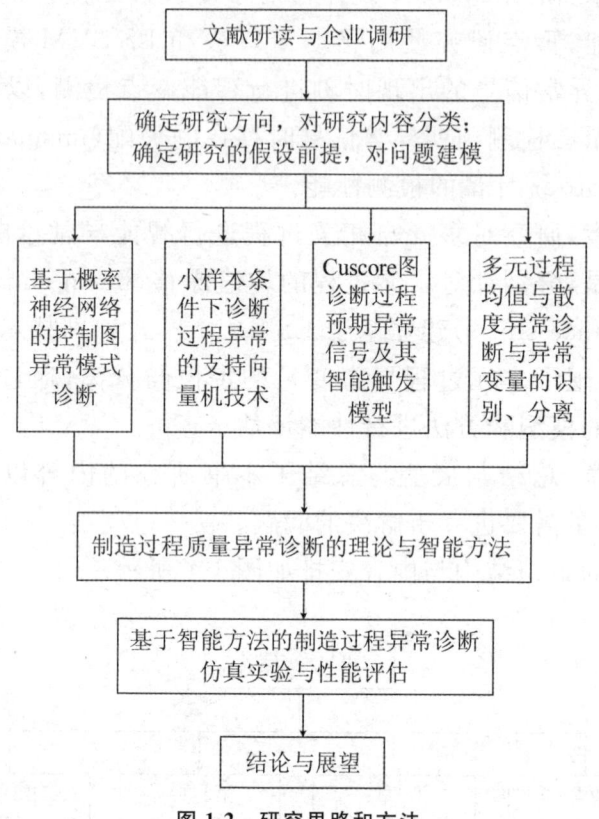

图 1-2 研究思路和方法

在本书研究的总体思路上,将统计学中的判断是否发生小概率异常事件的问题,转化为计算机科学中的模式识别问题来加以解决。在研究的方法和手段上,本研究主要通过计算机仿真实验来证明和验证提出的智能化技术方法的有效性,比如,通过蒙特卡罗仿真实验来模拟产品制造过程的生产数据,然后使用生成的仿真数据,通过 MATLAB 仿真软件对提出的智能诊断模型与算法进行编程仿真实验。另外,研究中的部分实验数据来自于其他公开发表的高水平国内外期刊论文。

第 2 章 基于 PNN 的制造过程质量诊断

神经网络是人工智能领域的重要研究成果,使用神经网络技术处理和挖掘过程中包含的质量信息,可以达到监控过程异常的效果。本章针对使用神经网络诊断过程中遇到的神经网络结构设计困难、参数选择与优化等问题,提出使用基于概率神经网络的控制图模式识别方法,对制造过程进行质量诊断,在简化神经网络诊断设计问题的同时,获得了很高的模式识别率。该方法适用于批量生产条件下的制造过程质量诊断。

2.1 过程异常与控制图的使用

2.1.1 两类过程波动理论与传统控制图

自从 1925 年休哈特建立统计过程控制(Statistical Process Control,SPC)理论,并针对过程的单一质量特性变量进行异常诊断以来,其所使用的主要工具就是 SPC 控制图(SPC Control Chart)。休哈特提出的统计过程控制理论认为,世界上没有两个产品是一样的,产品的质量特性之间存在的差异也称之为波动(Variation),波动是造成产品质量问题的根源,而产品质量特性数据的波动具有统计规律性。休哈特认为,造成产品数据波动的原因分为两类,普通原因(Common Cause)和特殊原因(Special Cause)。普通原因变差也称系统原因变差,是由过程中的随机因素造成的,这种随机因素在过程中总是存在,对过程质量的影响较小,只能通过管理手段来进一步减小,而不可能彻底消除;特殊原因变差是由过程中的异常因素造成的,对过程质量的影响较大,但是可以通过技术手段加以消除。休哈特将造成过程异常的原因称为"非随机原因"(assignable causes),戴明(Deming)将造成过程异常的原因称之为"特殊原因"(special causes)[138]。过程中一旦出现异常原因,就称过程处于失控(Out of Control)状态;当过程中不存在影响过程的特殊原因时,就称过程处于受控(In Control)状态。进行 SPC 质量管理的目的就是对过程中出现的异常波动加以迅速检测和诊断,然后采取纠正措施(Corrective Action)来排除造成过程异常波动的因素,使过程回归正常波动下的受控状态;否则,过程会因为存在异常波动而失控,导致大量废品的产生,给企业和客户造成巨大损失。

统计过程控制理论以传统统计学中的大数定理为理论基础,是大批量生产条件下制造业进行质量监控的产物。目前在批量生产中仍然是进行质量控制的有效方法,建立在小概率原理之上的休哈特 SPC 控制图被证明是诊断过程是否出现异常的有力工具。小概率原理认为:"小概率事件在一次实验中是不可能发生的,一旦发生,就认为过程中出现了非随机因素导致的过程异常。"休哈特在建立 SPC 控制图时使用了 3σ 准则,所谓的"3σ 准则"就是"对于满足正态分布的样本数据,分布在正负三西格玛范围的概率是 99.73%。样本数据落在正负三西格玛范围以外的概率是 0.27%,是个很小的概率,如果样本数据点落在正负三西格玛控制限之外,意味着发生了小概率事件,因此判定过程发生异常"。休哈特控制图的设计思想是先将使用控制图进行过程异常诊断犯第一类错误的概率 α 控制得很小(0.27%),然后使用其他规则来控制犯第二类错误的概率 β。由此产生了两类判异准则:

(1)样本点超过正负三西格玛范围,即对过程判异。

(2)样本点在正负三西格玛范围内出现非随机排列,即对过程判异。

满足上面第二条判异准则的数据点分布形式,在理论上可以找到很多种,但其中的很多种异常数据点分布形式在制造过程现场没有明显的物理意义。

休哈特博士在假设过程满足正态分布的前提下,基于统计学原理建立了诊断过程异常的 SPC 控制图方法,使得人们对于过程状态的判断建立在严格的统计学分析的基础之上,是质量管理与工程领域的重要成果,SPC 质量管理技术在现场的使用极大地提高了制造业的质量水平。

近年来,随着数据自动采集技术、计算机技术、网络技术、智能计算技术等新技术的出现,质量人员在传统 SPC 工具系列的基础上,在利用新技术建立智能化、自动化、实时质量诊断系统方面,做了大量探索性理论研究和推广应用工作,取得了一定的成果。

2.1.2 两个阶段的控制图及其使用

控制图可以在生产过程的两个阶段——分析阶段和控制阶段中使用。在分析阶段使用的控制图被称为分析用质量控制图,在控制阶段使用的控制图被称为控制用质量控制图。两阶段的控制图有不同的使用目的和方法。目前,对于使用基于统计学的控制图诊断过程异常的研究大多针对生产过程的控制阶段。本书在研究制造过程异常的诊断方法时,同样是研究当过程处于控制阶段时,如何使用人工智能的方法对过程异常加以诊断和异常变量的识别。

2.1.2.1 分析用控制图

在生产过程的分析阶段,首先需要采集过程数据,绘制控制图。控制图的作用是判定过程是否处于稳定状态,如果过程没有稳定受控,首先就要查找过程不稳定的原因并采取措施使得过程达到稳定受控状态。不稳定的过程不能投入生产,对不稳定的过程采取的纠正调整措施一般是技术措施,如改变工艺和参数等。

当过程已经稳定受控,就需要计算过程能力指数 C_{pk} 并判断其是否符合要求,如果 C_{pk} 达不到要求,就需要采取全面的管理措施,针对过程现场的"人员、机器设备、原材料、操作方法、测量和环境"等因素进行改善,直到过程能力指数 C_{pk} 满足要求为止,才能进入下面的生产控制阶段。可见,过程能力指数 C_{pk} 是诊断过程的能力是否满足了客户要求的一个量化指标。

过程能力指数 C_{pk} 体现的是过程的能力对于客户要求的满足程度。从管理的角度来看,过程能力指数 C_{pk} 是企业技术管理能力的体现,可以用来比较供应商之间以及同行业企业之间的质量水平;从统计学的角度来看,过程能力指数 C_{pk} 是评价过程变差的一个量化指标,反映的是稳定受控过程的客观现实的状态,但 C_{pk} 中包含了客户的规格要求这一主观因素。设客户的规格要求的上限和下限分别是 T_U 和 T_L,制造过程的均值和方差分别为 μ 和 σ,当假定过程输出的均值与规格中心重合时,则对于客户的"望目""望大"和"望小"三种规格情形的过程能力指数分别用 C_p、C_{pU}、C_{pL} 来表示,它们可以分别使用下面的公式计算:

$$C_p = \frac{\min\{T_U - \mu, \mu - T_L\}}{3\sigma} \quad (2.1)$$

$$C_{pU} = \frac{T_U - \mu}{3\sigma} \quad (2.2)$$

$$C_{pL} = \frac{\mu - T_L}{3\sigma} \quad (2.3)$$

在生产过程的分析阶段,当过程处于稳定受控状态并且过程能力满足要求时,另外一个重要的任务就是使用过程数据通过计算获得过程的参数,如,过程的均值、方差等,以便为生产控制阶段中使用的控制图提供控制限标准。

2.1.2.2 控制用控制图

在生产过程的控制阶段,需要按照一定的采样间隔采集过程数据并绘制控制图,当发现过程出现异常时,需要查找导致过程异常的原因和过程异常的部位,针对造成异常的原因采取技术措

第 2 章 基于 PNN 的制造过程质量诊断

施,使过程重新回归正常受控状态,并不断持续改进,使过程保持稳定受控的工作状态。

GB/T 4091—2001 规定了八条过程异常判定准则。在这些准则中,将控制图等分为 6 个区,如图 2-1 所示,每个区宽 σ,6 个区的标号分别为 1,2,3,3,2,1,则八条过程异常判定是:

① 1 个点落在 1 区以外。
② 连续 9 点落在中心线同一侧。
③ 连续 6 点递增或递减。
④ 连续 14 点中相邻点交替上下。
⑤ 连续 3 点中有 2 点落在中心线同一侧的 2 区以外。
⑥ 连续 5 点中有 4 点落在中心线同一侧的 3 区以外。
⑦ 连续 15 点落在中心线两侧的 3 区内。
⑧ 连续 8 点落在中心线两侧且无一在 3 区内。

上述八条判断过程异常的准则是在保持犯一类错误率 $\alpha=0.0027$ 的情况下得到的。GB/T 4091—2000 同时指出这八条过程异常判定准则只适用于(均值)控制图和 X(单值)控制图,并假设过程观测值服从正态分布。

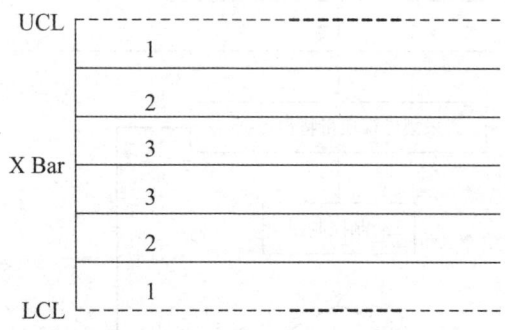

图 2-1 控制图判异准则示意图

两个阶段的控制图的使用情况可以用图 2-2 来总结。控制图实际是统计学显著性假设检验原理的一种图形化实现方法,它以图形的方式显示制造过程的质量特性数据随时间变化的波动情况,并判断过程是否受到异常因素的影响,发出异常报警,提醒人

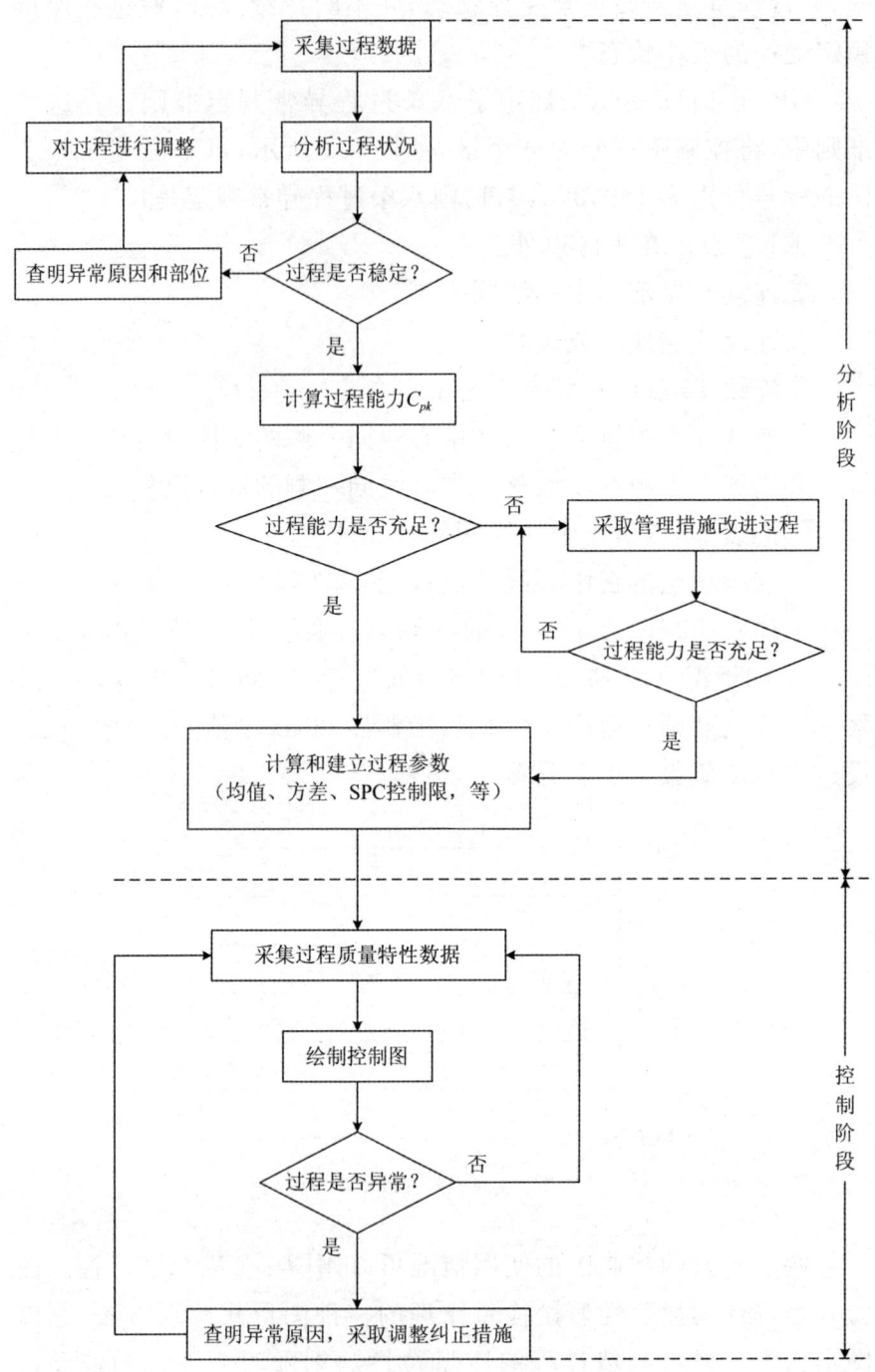

图 2-2 两阶段控制图使用流程图

们及时对过程做出调整,消除过程中的异常因素,保持过程处于稳定受控状态,最终保证产品和服务满足客户的要求。SPC是一整套的质量工具,除控制图外,还包括直方图、排列图、因果图、相关图、控制图、分层法和调查表等工具。另外,SPC不仅可以用于制造过程,而且也可以用于包括服务过程在内的一切管理过程。

2.2 控制图模式识别问题

2.2.1 控制图模式现象

统计过程控制图理论给出了判定过程异常的准则,然而,这些准则并不能对所有的过程失控都进行有效检测或诊断,比如,GB/T 4091—2001中的判异规则"连续14点中相邻点交替上下"定义了一种周期性异常,但对于控制图中出现的其他周期性异常没作定义,加上过程中偶然波动的影响,所以通过规则来描述所有周期性异常是无法做到的。现实生产过程中存在的另外一类重要的过程失控现象就是,过程失控在控制图中经常以异常模式的形式出现。由于过程中存在噪声,使用GB/T 4091—2001中规定的判断失控的准则却对此并不能进行有效的过程异常诊断。西方电气公司早在1958年就首先提出了控制图模式(Control Chart Pattern,CCP)现象,并对这一现象进行了研究总结和定义[85,86],AT&T公司在其AT&T SQC Hanbook[87,88]中对此也有相关描述,但这种类型的质量异常问题在当时的技术条件下没有得到有效解决。西方电气公司提出的基本控制图模式包括:正常模式,向上趋势模式,向下趋势模式,向上平移模式,向下平移模式,周期循环模式,其形状如图2-3所示。随着人工智能技术的发展,通过控制图的模式识别来诊断过程异常,成为过程质量诊断领域的研究内容之一。

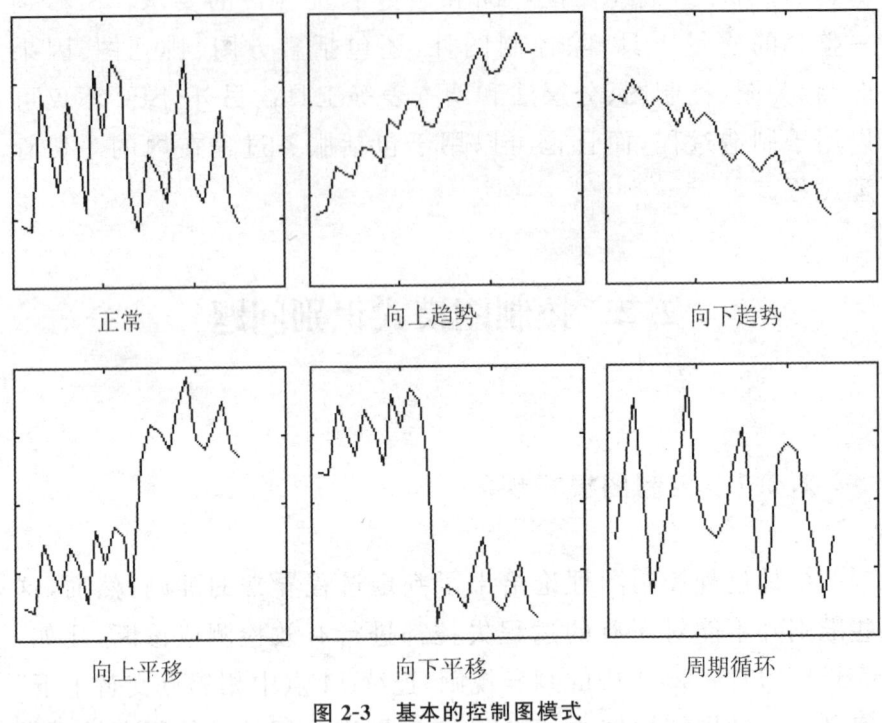

图 2-3 基本的控制图模式

实际应用中,存在着上述基本控制图模式的组合模式,比如,向上趋势+向下平移组合模式,向下趋势+循环组合模式等。不同的控制图组合模式反映了过程的不同状态和导致过程异常不同的原因。此外,对于某种控制图异常模式,还应当识别和估计其参数,如向上(下)平移模式的幅度、向上(下)趋势模式的斜率、循环模式的周期等。显然,对各种控制图模式的基本模式和组合异常模式进行诊断、特征识别和参数估计,同样是质量诊断研究工作的重要内容[89-92],通过对于这些问题的研究,可以为确定过程异常的发生原因和部位、为寻找和确定质量改进的措施提供更多的诊断信息。本书这一章的研究就是在生产过程的控制阶段,使用概率神经网络技术,对于无法使用统计学 SPC 控制图来进行判断的过程异常模式进行智能化识别,从而达到过程监控与异常诊断的目的。

2.2.2 描述控制图模式的 Monte-Carlo 方法

对于过程中控制图模式现象的研究,使用的数据最好是来自于过程现场的真实数据,这样可以真实地反映制造过程的实际情况,为质量诊断理论和技术的研究工作提供可靠的数据基础。在大批量制造环境下,企业可能积累了足够的过程历史数据,使用这些数据,会给质量诊断研究工作带来方便和益处。但是,很多情况下,企业并没有积累和存储质量诊断研究工作所需要的、大量的、各种类型的、反映了过程各种状态的样本测试数据。这种情况下,如果想要获得生产过程的真实数据以供研究使用,就会导致在生产设备、人员以及时间等方面的质量成本增加,在很多情形下受到各种条件的限制,很难实现。这时,研究工作可以使用数据仿真技术进行,并取得良好的效果。数据仿真技术是目前进行质量诊断研究时经常采取的手段。研究人员对于上述控制图模式数据的仿真描述一般使用蒙特卡罗(Monte-Carlo,MC)仿真方法[30,93]。相比于过程现场数据,仿真数据的使用显然会给研究工作带来很大的经济性。

2.2.2.1 *仿真数据的生成*

使用 Monte-Carlo 方法生成图 2-3 中所示的控制图模式数据所使用的公式如下:

设需要使用 Monte-Carlo 仿真方法产生的过程现场数据集为 $x(t)$,则

$$x(t) = \mu + d(t) + r(t) \tag{2.4}$$

其中,t 是过程观测数据的采样时刻,$x(t)$ 是 t 时刻过程数据的观测值。μ 是制造过程处于稳态受控状态时过程观测数据的均值。$r(t)$ 是过程的普通影响因素所造成的变差,服从均值为 0、方差为 σ^2 的正态分布,$r(t) \sim N(0,\sigma^2)$,即仿真是假定 $r(t)$ 为高斯白噪声。$d(t)$ 是由特殊原因造成的过程扰动;当过程中不存在异常原

因变差时,$d(t)$为0。不失一般性,本书的仿真研究中假设$\mu=0$,$\sigma=1$。显然,控制图模式的类型取决于$d(t)$,过程质量诊断的任务就是要对$d(t)$状态的变化情况进行辨识和判断。

1)对于正常控制图模式
$$x(t) = r(t) \tag{2.5}$$
其中,$r(t) \sim N(0,1), d(t) = 0$。

2)对于向上趋势控制图模式
$$x(t) = + k \times t + r(t) \tag{2.6}$$
其中,k是向上趋势模式的斜率,k的取值范围一般设定为$0.1\sigma \leqslant |k| \leqslant 0.3\sigma$。

3)对于向下趋势控制图模式
$$x(t) = - k \times t + r(t) \tag{2.7}$$
式(2.7)中k的意义同式(2.6)。

4)对于向上平移控制图模式
$$x(t) = + 1(t - t_0) \times s + r(t) \tag{2.8}$$

5)对于向下平移控制图模式
$$x(t) = - 1(t - t_0) \times s + r(t) \tag{2.9}$$
式(2.8)和式(2.9)中,s表示向上(下)平移的幅度,s的取值范围一般设定为$\sigma \leqslant |s| \leqslant 3\sigma$。$t_0$是一个决定平移发生时刻的参数。当$t \geqslant t_0$时,$1(t-t_0)=1$;当$t < t_0$时,$1(t-t_0)=0$。

6)对于循环控制图模式
$$x(t) = A \times \sin(2\pi t/T) + r(t) \tag{2.10}$$
式(2.10)中,A表示循环模式的幅度,A的取值范围一般设定为$\sigma \leqslant A \leqslant 3\sigma$。$T$是循环周期,$T$的取值范围一般设定为$2 \leqslant T \leqslant 8$。

2.2.2.2 仿真数据的标准化和编码处理

在生成用于训练神经网络和测试神经网络性能的仿真样本数据时,一般并不是直接将这些数据输入神经网络,通常要将仿真数据进行预处理,因为多数的神经网络只能接受一定范围的数据。预处理的方法包括规范化和编码。通过预处理,并没有改变训练样本数据和测试样本的本质特征,却可以简化数据表达的复杂度,提高神经网络对于数据的分类和识别能力,同时也可以提高神经网络在不同的生产条件下对于不同质量特性变量的样本数据进行处理的通用性。另外,经过训练的神经网络也只能接受与训练数据同样范围的测试数据,超出了这个范围,神经网络将失去其通过训练样本数据训练所得到的功能。

1) 数据标准化

数据标准化处理的目的是把各类量纲不统一的数据采用某种方式换算成具有可比性的数值,方法是通过变换将数据变化范围限定在某个固定的区间(如,[0,1]区间)。通常质量诊断研究工作中都假定制造过程满足正态分布,因此标准化变换公式为:

$$z(t) = (x(t) - \mu)/\sigma \qquad (2.11)$$

其中,$x(t)$ 为过程质量特性变量的观测值,μ 和 σ 的意义与式(2.4)相同。$z(t)$ 为 $x(t)$ 的标准化数值。对于实际的制造过程,其均值和方差需要使用过程观测数据通过估计才能得到,这时标准化公式变为:

$$z(t) = (x(t) - E(\mu))/E(\sigma) \qquad (2.12)$$

其中,$E(\mu)$ 和 $E(\sigma)$ 分别是当过程处于稳定受控状态时,由过程采样数据得到的过程均值和标准差的估计值。

2) 数据编码

经过规范化的数据,还可以进一步进行编码处理。对数据进行编码处理,实质是将过程的模拟观测数据进行数字化处理,以便去除数据中随机噪声对于神经网络训练的影响。这种噪声在过程现场被认为是来自于普通原因造成的波动。过程观测数据中噪声的存在,会使得神经网络的收敛速度变慢,影响使用神经网络技术进行过程异常诊断的效率。通过数据编码,有助于减少神经网络的训练时间。我们在使用神经网络技术诊断过程异常时,使用了如下的数据编码公式:

$$\begin{cases} y(t)=25, z(t) \geqslant 4.9 \\ y(t)=-25+N, -4.9+0.2(N-1) \leqslant z(t) < -4.9+0.2N \\ y(t)=-25, z(t) < -4.9 \end{cases}$$

(2.13)

其中，$y(t)$是对数据$z(t)$进行编码得到的数据。N是决定编码率大小的参数。对数据进行编码时,使用的编码率过小,将会导致数据特征丢失过多,不利于提高神经网络用于过程诊断时的分类能力；编码率过大,将导致训练神经网络时的运算量加大,同时也降低了神经网络的实时诊断能力。因此,必须合理选择编码率。本书在使用神经网络进行过程异常诊断时,取$N=1,2,\cdots,49$。

2.3　神经网络与控制图模式识别

　　质量诊断技术的发展是同生产制造技术的发展密切相关的,必须与市场对于生产的要求相适应,也必须与其他学科领域的最新技术成果相适应。随着数据采集技术、自动化技术、网络技术和人工智能技术的进步,使得原来在生产中需要人工来完成的许多工作可以而且必须由机器来完成,以实现生产过程状态的自动识别与诊断。特别是20世纪90年代以来,计算机集成制造系统(Computer Integrated Manufacturing System, CIMS)、敏捷制造(Agile Manufacturing, AM)、虚拟制造(Virtual Manufacturing, VM)等先进制造技术的出现,使得作为传统大批量生产的产物的SPC质量诊断技术,在很多方面难以适应新的制造模式下过程质量诊断与控制的要求。比如,传统SPC技术明显的缺陷有：

　　(1)传统SPC技术依赖于关于过程分布的假设条件,一般的研究都假定过程观测数据满足正态分布,但这一假设对很多过程并不成立,如连续性化工生产过程。

　　(2)控制图容纳的数据量有限,只反映了过程当前状态的信息,忽略了过程以往的历史数据信息和以往的过程诊断与调整经验。

市场环境和技术环境的变化,使得在制造过程中采用新的技术方法势在必行。神经网络(Neural Network,NN)作为典型的人工智能技术,对于过程观测数据具有学习、记忆、归纳、推理等智能化处理功能,可以在不依赖过程分布等精确数学描述的情况下,通过训练使其具有对过程状态进行分析和判断的能力,过程质量诊断是神经网络的一个重要的应用领域。

2.3.1 神经网络概述

2.3.1.1 神经元模型

人工神经网络技术[94-97]的发展最初来自于对人类大脑结构和运行机制的研究,最初的成果是建立了人工神经元模型。人工神经元是神经网络的基本非线性数据处理单元,是生物神经元的简化模型,可以用图 2-4 表示。

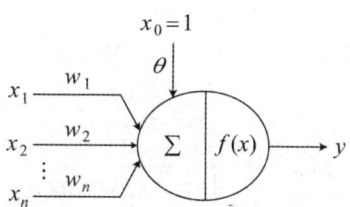

图 2-4 神经元结构模型

人工神经元的输入/输出关系可以表示为:

$$\begin{cases} I = \sum_{j=1}^{n} w_j x_j - \theta \\ y = f(I) \end{cases} \quad (2.14)$$

式(2.14)中,$x_j(j=1,2,\cdots,n)$ 是神经元的输入信号,该输入信号可以是来自神经网络以外的数据信号,也可以是来自其他神经元的信号;θ 为阈值。神经元有三个基本要素,一组连接,连接的强度使用权值来表示；一个求和单元，用于线性求取各输入信号的加权和；一个非线性激活函数，通过激活函数的非线性映射，将神经元的输出幅度限定在一定的范围之内（一般是[0,1]或者[-1,+1]之间）。

若把输入的维数增加一维，将$-\theta$看作是对应于输入$x_0 = 1$的权值，则式(2.14)变为：

$$\begin{cases} I = \sum_{j=0}^{n} w_j x_j \\ y = f(I) \end{cases} \quad (2.15)$$

其中，$w_0 = -\theta, x_0 = 1$。激活函数$f(\cdot)$可以有以下几种形式：

1) 阈值函数

$$f(x) = \begin{cases} 1, x \geqslant 0 \\ 0, x < 0 \end{cases} \quad (2.16)$$

2) 分段线性函数

$$f(x) = \begin{cases} 1, x \geqslant 1 \\ \frac{1}{2}(1+x), -1 < x < +1 \\ 0, x \leqslant -1 \end{cases} \quad (2.17)$$

3) Sigmoid 函数

常用的形式为：

$$f(x) = \frac{1}{1 + \exp(-\alpha x)} \quad (2.18)$$

其中，参数$\alpha > 0$，可以控制输出的斜率。另外一种是双曲正切函数

$$f(x) = \tanh\left(\frac{x}{2}\right) \quad (2.19)$$

在使用信号流图描述神经网络时，经常将神经元当成"节点"来看待。

2.3.1.2 神经网络的结构

神经网络就是由大量的神经元以一定的方式广泛连接起来的网络。神经网络的拓扑结构对神经网络的特征和性能有重要

影响。从总体结构上看,神经网络是三层网络结构:输入层、输出层和中间层。中间层又称为"隐层",可以由若干个隐层组成中间层。输入层和输出层分别包含多个输入节点和输出节点。中间层的每一层都包含多个节点。NN 的输入层和输出层的节点数目不是固定的,一般需要根据 NN 的应用场合和实际问题的特点来确定。NN 中间层层次的数目以及每一中间层的节点数目是个需要优化设计的问题,如果解决不当,将会降低使用神经网络解决问题的效果。

根据 NN 连接方式的特点,NN 主要分为两类:

1) 前馈型神经网络

前馈型 NN 的神经元包括两种:输入神经元节点和计算神经元节点。输入节点只是接收来自于神经网络外部的数据,可以有多个输出,作为其他神经元的输入;计算节点可以有多个输入,但只有一个输出值,该输出值可以耦合到其他多个节点作为其输入。前馈 NN 中的各节点只是接收前一层的节点的输出信号作为其输入,不接收来自其后面各层节点的信号,因此不存在反馈。

2) 互联型神经网络

互联型 NN 的组成节点全部是计算单元,同时也可作为输入/输出节点,向外界输入/输出信息。互联型 NN 的各个神经元之间都有可能存在连接。反馈型 NN 属于互联型网络。

2.3.1.3 神经网络学习的类型

神经网络需要首先通过向外界环境学习来获得知识,然后才具备解决问题的能力。神经网络的学习也被称为"训练",也就是按照某种既定的度量方式和学习算法调整 NN 的参数(权值系数、阈值)。NN 的学习方式主要包括监督学习、无监督学习和强化学习三个类型。

1) 监督学习

监督学习需要同时给出输入和与该输入相对应的期望输出,期望的输出被称为"教师信号",教师信号是监督学习时评价学习效果的标准。当 NN 的输出与期望的输出有误差时,NN 将通过自身的调节机制调整相应的参数,使 NN 的输出向误差减小的方向变化,经过若干次的训练调整,当 NN 的输出与期望的输出相一致(或者两者的误差足够小)时,NN 完成对自身的训练。典型的监督学习算法是误差反向传播算法(Back Propagation,BP)。监督学习也称为有教师学习。

2)无监督学习

无监督学习不需要为 NN 提供外部教师信号,NN 根据外部环境提供的数据的某些统计规律,以自组织的方式调整自身的参数和结构,以获得外部输入信号的某些特征,网络的学习效果评价标准隐含于神经网络内部。无监督学习也称为无教师学习。

3)强化学习

强化学习方式中,外部环境同样不提供教师信号,但提供"奖/罚"激励信号,NN 学习系统通过强化受奖励的动作来调节自身的参数,提高自身的性能。强化学习虽然不提供对应于输入的目标输出,但提供输入的级别,以评价输入序列的性能。强化学习也称为再励学习。

对于平稳的学习环境,其统计特征不随时间变化,理论上认为采用监督学习的方式可以使神经网络记忆这些统计特征;对于非平稳的学习环境,其统计特征随时间变化,采用监督学习的方式通常无法跟踪这种变化,需要使用具有自组织、自适应能力的神经网络才能跟踪这种变化。

2.3.1.4 神经网络的学习规则

实现神经网络学习的具体算法被称为神经网络的学习规则,NN 根据这些规则来调整神经元之间的连接权重。

1)误差纠正学习规则(Delta 学习规则)

误差纠正学习的目标是使 NN 的期望输出与其实际输出之间的误差最小,常使用的目标函数是均方误差判据,思路是基于输出方差最小。常用的实现方法是最陡梯度下降法。BP 神经网络和模拟退火算法都使用该学习规则。

2) Hebb 学习规则

该学习规则由神经心理学家 Donall Hebb 提出。Hebb 规则来自于生理学中的条件反射机理,其内容为:当某个连接两端的神经元同时处于激活/抑制状态(激活同步),该连接的强度应该加强;反之,则应减弱。Hebb 规则可用于无教师学习,也可以用于有教师学习,具有自组织的学习能力。Hopfield 网络使用的是 Hebb 规则。

3) 相近学习规则

该规则对于连接权值的调整取决于神经元之间输出的相似性,如果两个神经元的输出比较相似,则它们之间的连接加强,反之,则它们之间的连接强度减弱。竞争型 NN 多采用这种学习规则,如 ART(Adaptive Resonance Theory)网络和 SOFM(Self-Organizing Feature Maps)网络中就采取该学习规则。相似学习规则拓宽了 NN 在模式识别与分类方面的应用。

神经网络的结构和学习方式决定了神经网络的性能和特点,必须根据实际问题的特点合理甄别和选择,以便达到好的应用效果。

神经网络是人工智能的重要组成部分,在信息处理方面具有突出的优点,比如 NN 具有很强的复杂非线性关系逼近和跟踪预测能力,对数据可以进行并行分布处理、容错性强,对未知的或不确定的系统具有学习和自适应能力,等等。神经网络在模式识别与图像处理、控制和优化、预测预报与信息智能处理、过程与设备的故障诊断等众多领域取得了成功应用。

2.3.2 控制图模式识别的神经网络方法

控制图模式识别是基于 SPC 的过程质量智能诊断的基础,大量的过程智能诊断研究工作都是针对这一问题展开的[98-104]。人工神经网络应用于过程异常诊断的优势在于,ANN 技术不需要被处理的观测数据满足正态性,对于难以使用统计学描述的连续流程性工艺过程,同样可以使用经过训练的 ANN 进行分析处理;对于过程参数变化的过程,综合运用数据采集和计算机技术,可以通过在线的方式训练神经网络,对过程的状态信息进行学习、辨识和分类处理,ANN 具有的自组织、自适应能力有利于对过程的复杂异常状态作出合理判断,ANN 技术大大拓展了以往基于统计学 SPC 诊断技术的功能。但是,使用 ANN 技术来诊断过程的异常状态在技术上存在一定问题,即 ANN 识别器的结构设计问题,包括如何合理确定神经网络隐层的数目以及每个隐层中节点的数目。以往的研究对于这个问题的处理都是采用个人经验或者经过简单的试验设计来解决的,不仅费时多、效率低,而且识别效果不理想,模式识别率不高,比如,BP 神经网络是在控制图模式识别的研究中广泛使用的网络,在 Guh[85,103] 的研究中,使用 BP 神经网络对六类控制图模式进行识别,其 BP 网络结构依靠个人经验反复选择后,BP 网络对于每类模式的识别率很多在 90% 以下,显然需要进一步提高,否则在 ANN 技术实际应用于过程诊断时会导致较大的误判,导致较大的质量损失。

针对使用 ANN 进行控制图模式识别中存在的网络结构设计困难、对控制图模式的识别率不高的问题,本章下面的工作对此加以研究解决。在考察各类神经网络结构和性能的基础上,提出使用概率神经网络(Probabilistic Neural Network,PNN)进行控制图的模式识别,在简化神经网络设计的同时,提高神经网络的训练速度,提高控制图模式识别率及过程异常诊断的工作效率。具体研究内容包括:设计 PNN 的结构,选择并优化其参数,通过仿真实验研究,从识别率和 ARL 两个方面评价提出的 PNN 方法对于控制图异常的诊断性能。

2.4　基于 PNN 控制图模式识别的过程异常诊断

对于过程异常现象进行诊断,其实质是对过程的状态进行识别或者分类。概率神经网络[105-107]就是一种适用于模式分类的径向基神经网络。PNN 最初由 Specht 提出,其算法等同于使用前馈并行算法、无局部最小化的贝叶斯最优分类器。PNN 的学习算法在概率上依据贝叶斯准则收敛,算法简单,PNN 的隐层不需要设计,不会因为不同的 NN 结构设计问题而影响到分类的效果,在应用于工业过程的质量诊断时,易于在不同的制造过程之间形成可以比较质量水平的工业质量技术规范。

2.4.1 概率神经网络原理

PNN 原理上利用样本的先验概率和 Bayes 决策理论对新的样本进行分类,PNN 具有统计分类的功能,但并不受正态分布等条件的限制,在其运算过程中可读出被分类的新的输入样本的后验概率,使得分类结果得到了一定的概率方面的解释,这是 PNN 在应用于过程质量诊断时的一个优点;PNN 进行模式分类时使用了自监督的前馈网络分类方法,无须训练网络的连接权值,由给定的训练样本直接构成隐层单元就可以检验网络的性能。PNN 在分类功能上等同于最优 Bayes 分类器,其分类能力优于 BP 网络。PNN 不像其他前向网络那样需要使用 BP 算法进行误差反向传播计算,而是完全前向的计算过程,相比于 BP 算法网络,具有训练时间短且不易收敛到局部最优点的优点,易于发展成为可以在线工作的神经网络。同时,PNN 网络的结构简单,使用时易于设计和实现,减少了神经网络的设计时间,在过程诊断等实际应用中具有推广价值。

2.4.1.1 Bayes 分类器

PNN 利用贝叶斯分类器方法和非参数概率密度函数的 Parzen 窗估计方法[108,109],对模式识别问题给出了通用的解法。

Bayes 分类器的设计思想主要是源于 Bayes 决策理论。设有 $\omega_i (i=1,2,\cdots,c)$ 个待匹配的类模式集合,每个类的先验概率为

$P(\omega_i)$,在没有先验知识的情况下 $P(\omega_i)$ 通常取相等的概率值。对于任意的一个随机的矢量样本 $X \in R^d$,每个类的类条件概率密度函数为 $P(X|\omega_i)$,d 为训练样本矢量的维数。设 L_{ji} 为 ω_i 被判断为 ω_j 所担负的风险因子(期望损失),L_{ji} 可以依据不同决策情况根据专家经验做出假设,在没有专家经验的情况下,$\forall j \neq i = 1, 2, \cdots c$,可以认为 $L_{ji} = L_{ij}$。

按照 Bayes 最小平均条件风险决策准则,$\forall j \neq i = 1, 2, \cdots c$,可以对未知模式的样本 X 使用如下的诊断规则进行判决:

$$X \in \omega_i, \text{若有 } r_i(X) < r_j(X) \tag{2.20}$$

式(2.20)中,$r_j(X)$ 表示将样本 X 判决属于 ω_j 类模式所造成的风险损失。

$$r_j(X) = \sum_{i=1}^{c} L_{ji} P(X|\omega_i) P(\omega_i) \tag{2.21}$$

其中,$j = 1, 2, \cdots, c$。式(2.21)中的关键在于类条件概率密度 $P(X|\omega_i)$ 如何求取,因为一般情况下它是未知的,可以使用 Parzen 窗法从已知随机样本中估计概率密度函数,只要样本数目充足,由该方法所获得的函数可以连续平滑地逼近复杂的原概率密度函数。

设属于模式 ω_i 的训练样本 $X^{(i)}$ 的数目为 N_i,其中 $i = 1, 2, \cdots c$;X 为待匹配样本;$X^{(i)}$ 在概率神经网络中作为权值。Parzen 窗函数取高斯核函数,由 Parzen 窗函数法可以得到模式 ω_i 的类条件概率密度估计为:

$$\hat{P}(X|\omega_i) = \frac{1}{(2\pi)^{d/2} \sigma^d} \cdot \frac{1}{N_i} \sum_{i=1}^{N_i} \exp[-(X - X^{(i)})^T (X - X^{(i)})/2\sigma^2] \tag{2.22}$$

其中,σ 为 Parzen 窗的窗宽度(平滑参数因子),其取值确定了以样本点为中心的钟状曲线的宽度,并且决定了模式样本点之间的影响程度,关系到概率密度分布函数的变化。显然,$\hat{P}(X|\omega_i)$ 是多元高斯分布在每个训练样本处的和。这里的核函数不仅仅限于高斯函数,还可以采用其他形式的核函数。

2.4.1.2 概率神经网络的结构

概率神经网络(PNN)属于径向基神经网络的一个拓展分支,采用前馈网络结构,采用有监督学习方式。PNN 基于概率统计思想进行模式分类识别,训练速度快且避免了局部最小化。

PNN 通过 Bayes 分类规则与 Parzen 窗方法来实现模式分类,其典型的神经网络模型结构共分为四层[110]:输入层、模式层(又称为样本层)、求和层、输出层(又称为竞争输出层,决策输出层),如图 2-5 所示。

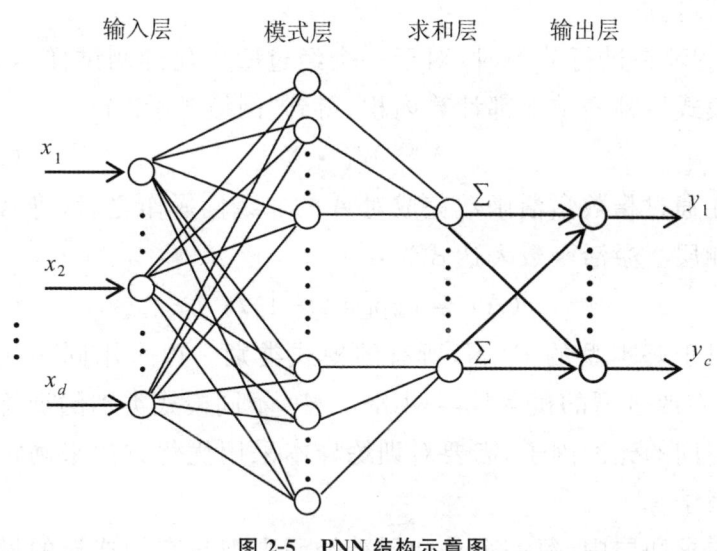

图 2-5 PNN 结构示意图

图 2-5 中,网络的输入矢量为 $X = [x_1, x_2, \cdots, x_d]^T$,其输出为 $Y = [y_1, y_2, \cdots, y_c]^T$。

1) 输入层

PNN 的输入层神经元的数目等于训练样本输入向量的维数 d。输入层各神经元不做任何计算工作,只是简单地直接将输入变量传递给模式层。

2) 模式层

模式层神经元的个数为各个类别的所有训练样本数之和,并

被分为 c 类。

假设我们随机地从 c 个类别的样本中选取 n 个 d 维的训练样本 $X_k, k=1,2,\cdots,n$。PNN 学习过程为：首先，训练样本数据集中的每一个样本都被归一化为单位长度；然后，初始化从输入层到模式层的权值，$W_k = X_k$。其中，W_k 为所有输入层神经元至第 k 个模式层神经元的连接权值向量。这样得到的 PNN 网络，输入层各神经元与模式层各神经元之间是完全连通的，而模式层各神经元只与同它相对应的求和层神经元相连接。可见，PNN 的训练实现方式简单，只要将各类的训练样本作为输入层与模式层的权值即可。

PNN 在进行分类时，对于一个经过归一化的测试样本 X，每一个模式层神经单元都计算内积，得到"网络激励"A，

$$A = W_k^T \cdot X \quad (2.23)$$

在通过指数激活函数完成对 A 的非线性操作之后，将 A 输出到求和层。激活函数表达式为：

$$f(A) = \exp[(A-1)/\sigma^2] \quad (2.24)$$

对于基本型的 PNN，所有的模式类别一律采用同一光滑因子 σ。在改进型的概率神经网络[111]中，不同模式类别的训练样本采用不同的光滑因子，需要对训练样本采用优化方法来确定各个光滑因子 σ。

在求和层中，每一个求和层神经元则把与它相连接的模式层神经元的输出结果进行相加。

3）求和层

求和层的神经元与待匹配的模式类别一一对应，求和层的神经元将对应的模式层的一组神经元输出求和，从而得到各个类别模式的类估计概率密度函数。不难看出，网络的模式层和求和层的作用就是为了实现贝叶斯决策方法。

4）输出层

决策输出层节点数目等于待匹配的模式类别数目，采用 Bayes 分类规则和竞争算法，在各个模式的类估计概率密度中选

第 2 章 基于 PNN 的制造过程质量诊断

择一个具有最大后验概率密度(即具有最小"风险")的神经元作为整个系统的输出。

当训练样本的数量增加时,PNN 的模式层神经元数目将随之增加;而当数据的模式类别增加时,求和层和输出层神经元将增加,所以,随着过程先验知识的积累,概率神经网络可以不断横向扩展,其对过程状态异常和故障的诊断能力将不断提高。

2.4.1.3 概率神经网络的特性

PNN 在应用于制造过程异常模式诊断时,与以往研究中使用较多的 BP 神经网络相比,具有很多优良特性,主要体现在以下几个方面:

1)设计简单、收敛速度快,易于工程实现

在进行网络设计时,BP 网络的输入/输出层和 PNN 相同,但 BP 网络隐层的数目和每一隐层神经元的数目的设计没有确定性法则,需要根据设计者的个人经验反复试验选取,这样可能导致不同的神经网络的设计者对于过程异常诊断的结果相差很大。而 PNN 需设计和调节的参数少,不需要设计隐层(模式层和求和层)的结构,比较容易掌握和使用。BP 网络使用反向误差传播学习算法,算法的收敛速度慢,而且易陷入局部极小值。PNN 的学习过程简单,将训练样本归一化为单位矢量后即可直接输入到网络中,训练网络所用时间仅仅略大于网络读取输入数据的时间,且不会陷于局部最优解。这一特点有利于在生产过程现场以实时的方式对 PNN 进行训练和对过程进行实时在线诊断。

2)PNN 在算法上依据贝叶斯准则收敛,诊断结果稳定性高

BP 网络的训练结果受连接权值的初始值设置影响较大,训练结果会随初始权值设置的不同而不同,这样可能导致分类结果存在比较大的差异,从质量诊断的角度来讲,使用 BP 网络对过程进行诊断,给出的是一个不确定、不便于解释的结果。PNN 对过程进行诊断时,不仅分类准则的概率意义明确,而且可以最大程度地利用过程制造的先验知识,在大样本条件下,总可以保证获

得基于贝叶斯准则下的最优解,因此,使用PNN对不同过程进行异常诊断的结果,易于比较不同过程的质量水平。

3) 样本的更新能力强,适用于构建在线实时检测系统

过程的观测数据需要不断更新才能反映变化了的过程状态。如果在过程质量诊断过程中需要加入新的训练样本或需要淘汰某些旧的训练样本,从PNN使用者的角度来看,PNN的隐层的变化是不需要考虑的,不需要增加额外的训练工作,PNN对于新样本的适应和学习的时间仅仅相当于或略大于PNN对新样本的读取时间,使得PNN易于在线工作。而对于BP网络,在训练样本发生变化后,则需要对BP网络的权值全部重新训练,增加了工作负荷和时间消耗。

2.4.2 PNN控制图模式识别的仿真实验

对于过程质量异常现象,需要使用迅速、高效的技术手段加以分析、诊断和识别,以便对生产过程的质量问题进行调整和处理,使生产过程处于稳定受控的状态。控制图是质量监控和诊断的重要工具,本书这部分的研究内容使用概率神经网络技术、以控制图的模式识别的方式,达到过程质量诊断的目的。本书对使用PNN技术进行控制图模式识别的研究采用计算机仿真的方式进行,使用的仿真工具是MATLAB软件。仿真实验过程如下:首先是生成概率神经网络PNN,生成训练样本数据和测试数据,并进行预处理;然后,使用训练数据对PNN进行训练;再然后,使用测试数据对PNN进行测试并优化选择PNN的参数;最后,对设计好的PNN模式识别器进行性能评价。

2.4.2.1 PNN设计

相对于BP等常用的神经网络,PNN在进行模式分类时的优势之一就是其设计工作简单、高效,只需要合理确定训练样本矢量的维数和模式类别数,即可完成PNN的结构设计。我们的研

第 2 章 基于 PNN 的制造过程质量诊断

究对 PNN 进行了如下几个方面的设计。

1)输入层设计

PNN 的输入层主要是设计输入层神经元的个数,需要根据具体的问题来确定。对于控制图模式识别问题,PNN 输入层神经元的数目,应当等同于一张 SPC 控制图中包含的数据点的数目(即 SPC 控制图窗口的宽度)。在控制图使用中,如果控制图中的数据点数目少,检测过程异常的速度快,但对过程异常的误判率会增高,第一类错误增大;如果控制图中数据点的数目多,检测过程异常的速度慢,但对过程异常的漏判率会降低,第二类错误增大。本书研究中采用了美国汽车工业行动集团(Automotive Industry Action Group,AIAG)的 TS 16949 系列质量技术标准中 SPC 手册的控制图标准[112],将 PNN 的输入层神经元的数目设计为 25。

2)模式层和求和层设计

PNN 的隐层(模式层和求和层)在原理上不需要由使用者设计,在仿真实验中由仿真软件自动生成。

3)输出层设计

PNN 输出层的神经元数目应当等于需要识别的样本数据类别的数目,研究中涉及六类控制图,将 PNN 输出层的神经元数目设计为 6。

4)参数设计

研究中,典型的 PNN 需要设计和确定的参数只有高斯核函数的宽度 σ,该参数对应于 MATLAB 仿真软件的函数"newpnn"参数"SPREAD"。函数"newpnn"在 MATLAB 仿真软件中用于生成概率神经网络。

根据上述设计思路和原则,使用 MATLAB 仿真软件进行了 PNN 的设计,得到的 PNN 结构图见图 2-6。

2.4.2.2 PNN 的训练和测试

与其他神经网络一样,在使用 PNN 对某个控制图的模式进行识别之前,必须使用训练样本数据对其进行训练,然后测试PNN的识别效果。

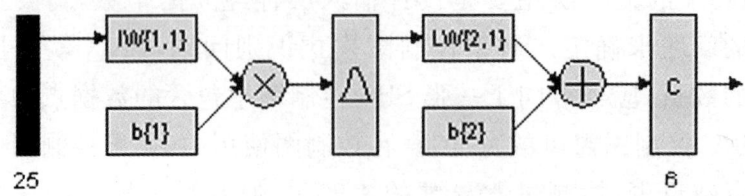

图 2-6　仿真实验中设计的 PNN 结构图

1)训练/测试样本的生成

仿真实验中,使用本书 2.2.2 节中的方法,为六类控制图模式生成了样本矢量数据,每个样本数据包含 25 个数据点,组成一个样本矢量。各类控制图包含的样本矢量的数量是相同的,共生成 4000×6 个样本矢量。其中 3000×6 个样本数据用于训练 PNN,1000×6 个样本数据用于测试 PNN 的识别率性能。同时,由于 PNN 采取的是有监督的学习方式,对应于六类控制图模式(依次是:正常模式、向上趋势模式、向下趋势模式、向上平移模式、向下平移模式、周期循环模式),PNN 的输出矢量被分别定义为[1,0,0,0,0,0],[0,1,0,0,0,0],[0,0,1,0,0,0],[0,0,0,1,0,0],[0,0,0,0,1,0],[0,0,0,0,0,1]。生成的仿真样本数据经过数据预处理,并被归一化成为单位矢量。

在使用本书 2.2.2 节公式(2.4)～公式(2.10)生成仿真样本数据时,公式中各个参数设置情况归纳于表 2-1 中。

表 2-1　仿真实验样本数据的参数取值状况

参数	参数取值范围	参数取值说明
μ	0	稳态过程的均值,取为常数
σ	1	稳态过程的标准差,取为常数
k	[0.1,0.3]	向上/下趋势控制图模式的斜率,是在[0.1,0.3]范围内按照均匀分布生成的随机数

续表

参数	参数取值范围	参数取值说明
t_0	12,13,14	向上/下平移控制图模式的平移发生的时刻,在仿真实验中使用了25个点的控制图,t_0取值为随机整数,在12、13、14中随机取一个数
s	[1,3]	向上/下平移控制图模式的平移幅度,是在[1,3]范围内按照均匀分布生成的随机数
A	[1,3]	循环模式的幅度,是在[1,3]范围内按照均匀分布生成的随机数
T	8	循环模式的周期,取为常数

2) PNN的训练过程

仿真实验中,PNN的训练过程简单迅速,只是将训练样本集中的各类样本数据全部输入到PNN中,成为模式层各个神经元节点,即可完成训练。仿真实验采用的是离线训练、调整和测试的方式。对于实际的过程诊断问题,当神经网络完成训练和性能测试后,可以将神经网络以在线的方式应用于过程现场,对来自现场的控制图数据进行模式识别。

3) 参数调整

完成训练后,为了得到好的模式识别效果,需要调整和设计PNN的高斯核函数参数"SPREAD"的取值,该参数的取值对PNN控制图模式识别的效果有很大影响。当PNN得到较为理想的识别率后,才可以将其应用于过程的诊断。如果在PNN对于控制图模式识别率不高的情况下将其应用于过程诊断,会造成比较大的诊断错误(误判和漏判)和质量损失,给过程改进造成困难。

在对核函数参数"SPREAD"进行选择时,将其从0.5到16(以0.25、0.5或1为增量间隔)取值。最后,将参数"SPREAD"的取值选定为9,获得了较高的识别率。实验中参数选择的详细信息归纳于表2-2中。

表 2-2 仿真实验中"SPREAD"参数的选择情况

模式类型 SPREAD	正常模式	向上趋势模式	向下趋势模式	向上平移模式	向下平移模式	周期循环模式
0.5	0.999	0.031	0.029	0.001	0.001	0.001
0.75	0.964	0.722	0.725	0.502	0.492	0.591
1	0.949	0.937	0.933	0.901	0.902	0.963
1.25	0.951	0.939	0.927	0.896	0.899	0.958
1.5	0.940	0.933	0.933	0.902	0.891	0.961
2.00	0.946	0.931	0.931	0.898	0.897	0.961
3	0.945	0.937	0.935	0.897	0.894	0.961
4	0.955	0.936	0.935	0.909	0.900	0.965
5	0.958	0.936	0.938	0.905	0.911	0.967
6	0.976	0.935	0.938	0.924	0.923	0.966
7	0.981	0.944	0.943	0.934	0.933	0.965
8	0.989	0.940	0.939	0.942	0.941	0.969
8.5	0.993	0.937	0.943	0.946	0.945	0.967
9	0.995	0.940	0.936	0.946	0.951	0.967
9.5	0.995	0.932	0.932	0.946	0.9488	0.964
10	0.996	0.932	0.930	0.946	0.943	0.957
11	0.997	0.922	0.921	0.947	0.943	0.960
12	0.999	0.919	0.912	0.943	0.942	0.954
13	0.998	0.899	0.906	0.938	0.934	0.952
14	0.999	0.900	0.900	0.937	0.934	0.950
15	0.998	0.889	0.890	0.929	0.934	0.947
16	0.999	0.877	0.879	0.929	0.930	0.948

仿真实验中,PNN需要选择取值的只有"SPREAD"一个参数,本书的研究对此问题采用反复试验的"试凑"方法加以解决,其效果依赖于个人的经验。如果需要提高PNN对于控制图的识别率,就需要以更小的增量来刻画"SPREAD"参数的取值,加大

设计的工作量。特别是,当遇到设计的模式识别器有多个参数需要组合起来加以选择的情况时,使用"试凑"的方法选择和确定模式识别器的参数将更加困难,这种情况下可以考虑使用进化优化算法来确定识别器的参数,达到较为理想的模式识别效果。

2.4.2.3 PNN 的训练和测试

整个仿真实验的过程和步骤使用流程图来加以简单总结和说明,如图 2-7 所示。

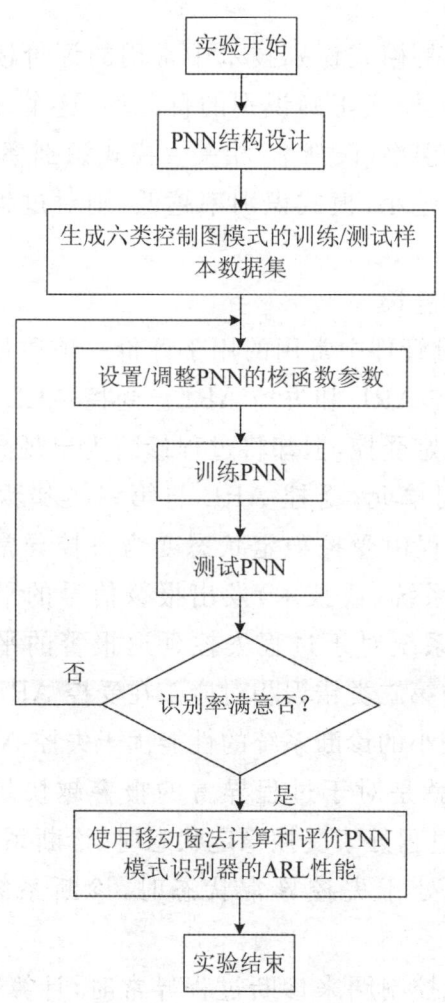

图 2-7 PNN 控制图模式识别实验过程

2.4.3 PNN控制图模式识别的性能评价

2.4.3.1 性能评价方法

仿真实验从两个方面来评价所采用的PNN控制图模式识别技术对于过程诊断的效果:模式识别率和失控平均链长(Average Run Length,ARL)。

1)模式识别率

模式识别率是模式识别技术中常用的评价模式识别技术的指标,表示将某类模式正确识别的百分率,这个评价指标与统计学中的第一类错误率(误判率)相关。模式识别率越高,则对过程诊断的误判率就越小;模式识别率越低,则对过程诊断的误判率就越大。

2)失控平均链长

ARL是质量管理中常用的用于评价一项质量诊断技术性能的指标,包括受控ARL和失控ARL。受控ARL是指,过程的实际工作状态为稳定受控,但却将过程诊断为出现异常(误发报警)的平均采样时间单元,受控ARL与第一类错误相对应。失控ARL是指,从过程由受控稳定状态变为失控异常状态的那一刻算起,直到诊断系统(或技术)发出报警信号的平均采样时间单元,也就是诊断系统对于过程失控延迟报警的平均采样时间单元,失控ARL与第二类错误相对应。在受控ARL的相同的情况下,失控ARL较小的诊断系统的性能优于失控ARL较大的诊断系统的性能,也就是对于过程异常的报警越快越好。理想情况下,人们希望当过程处于受控稳定状态时,诊断系统的ARL越大越好;而当过程处于失控异常状态时,诊断系统的ARL越小越好。

在使用传统控制图来诊断过程异常时,计算ARL使用的是统计学的方法[115,116]。在使用神经网络等智能技术研究过程异常诊断问题时,需要根据 ARL 的定义来加以计算。本书仿真研究中使用"移动窗"法来计算失控 ARL,以此来评价 PNN 模式识别器对于过程异常的诊断性能。"移动窗"的窗口长度就是一张控制图中包含的数据点数,根据美国汽车工业行动集团 SPC 手册,取窗口长度为 25。该方法每次给窗口中添加一个新的数据(同时剔除窗口中最后一个旧的数据),然后将更新的控制图模式数据提供给 PNN 模式识别器进行模式的分类,从而判断过程中是否存在异常因素(即判断控制图的异常模式是否出现)。使用移动窗法计算 ARL 的基本过程和原理如图 2-8 所示。

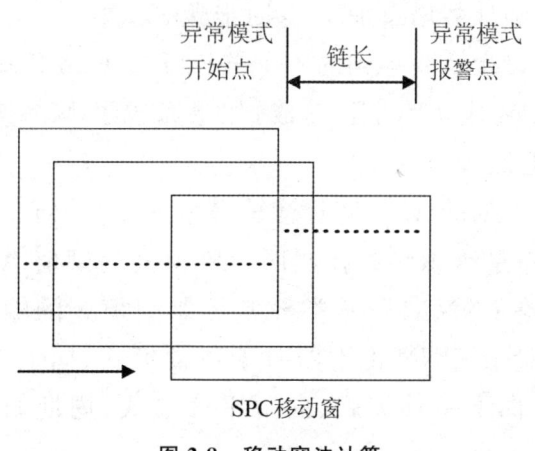

图 2-8 移动窗法计算

在计算 ARL 的开始时刻,提供给 PNN 模式识别器的是包含 25 个时间序列数据点的一张正常模式的控制图数据,然后,移动窗口每次向前移动一步。移动窗口每移动一步,就在移动窗口中加入一个新的属于某类异常模式的数据点(同时去除原来窗口中的一个旧的数据点),组成一张新的控制图时间序列数据,提供给 PNN 模式识别器进行识别,直到 PNN 识别器发出模式异常报警信号。链长就是从有异常模式数据进入移动窗口的时刻算起,到 PNN 识别器发出异常报警信号的时刻之间的时间间隔。

使用移动窗口法计算 PNN 识别器识别某类异常控制图模式的性能 ARL 的具体算法和步骤如下。

(1) 设定需要反复进行的实验次数 M。

(2) 为进行该类异常控制图模式是否出现的判断设置一个阈值 ARL_θ，将初始采样时刻设置为 25。

(3) 取出该类异常控制图模式的最近时刻的 25 个仿真数据 $X_j(j=t-24, t-23, \cdots, t-1, t)$，并将 X_j 进行预处理(规范化、编码)和单位长度处理，成为 $\overline{X}_j$。

(4) 将 $\overline{X}_j$ 组成矢量 $V_i = (\overline{X}_{t-24}, \overline{X}_{t-23}, \cdots, \overline{X}_{t-1}, \overline{X}_t)$，并送入 PNN 识别器进行识别，得到识别输出的结果 ARL_o。

(5) 如果 $ARL_o > ARL_\theta$，得出已经检测到异常模式的结论，转入下一步；否则计数增量加 1，返回步骤 3)。

(6) 记录该次实验得到的一个链长(RL, Run Length)。

(7) 判断 M 次实验是否完成，如果完成了 M 次实验，转入下一步；否则，返回步骤 3)。

(8) 通过得到的 M 个链长计算得出 ARL 的值。

上面判断是否识别到控制图异常模式的阈值 ARL_θ，是根据概率神经网络 PNN 对于某类异常控制图模式的输出值来设置的。仿真实验中，PNN 识别器对于正常模式、向上趋势模式、向下趋势模式、向上平移模式、向下平移模式、周期循环模式的输出矢量为 [1,0,0,0,0,0],[0,1,0,0,0,0],[0,0,1,0,0,0],[0,0,0,1,0,0],[0,0,0,0,1,0],[0,0,0,0,0,1]，因此，使用十进制表示的向上趋势模式、向下趋势模式、向上平移模式、向下平移模式、周期循环模式的阈值 ARL_θ 可以分别设定为 15.5, 7.5, 3.5, 1.5, 0.5。

2.4.3.2 仿真实验结果

1) 模式识别率

仿真实验中，当参数"SPREAD"的值为 9 时，对于完成训练的 PNN，对每类控制图使用 1000 个测试样本进行测试，得到了每类控制图模式的识别率，并将该识别率结果与 Guh(1999, 2002) 的研究结果相比较。仿真实验的结果如表 2-3 所示。

表 2-3　六类控制图模式识别的仿真实验结果（识别率）

识别率 \ 模式类型	正常模式	向上趋势模式	向下趋势模式	向上平移模式	向下平移模式	周期循环模式
本书结果	0.995	0.940	0.936	0.946	0.951	0.967
Guh(1999)的结果[100]		0.86	0.86	0.90	0.90	0.92
Guh(2002)的结果[85]		0.86	0.89	0.90	0.89	0.94

从表 2-3 中可以看出，本书仿真实验使用 PNN 识别器对六类控制图模式的识别率全部高于 93%，该结果明显好于 Guh 的研究结果。同时可以看到，使用相同的实验条件（对各类模式的控制图的训练样本数量相同，参数设置相同），PNN 模式识别器对各类控制图的识别性能并不相同，这种现象同各类控制图之间的特征是否有明显的区别密切相关。实验过程中通过进一步观察，发现一定数量的向上/下趋势模式被错误判断成向上/下平移模式，而一定数量的向上/下平移模式被错误判断成向上/下趋势模式，影响到识别率的进一步提高。鉴于此，为了进一步提高控制图模式识别率，可以考虑综合使用其他智能技术（如小波[113,114]技术）来进一步展开研究。

2) ARL 性能

仿真实验分别对向上趋势模式、向下趋势模式、向上平移模式、向下平移模式、周期循环模式进行 1000 次（$M=1000$）试验，并记录每次实验的链长，最终得到各类异常模式的 ARL，仿真试验的结果总结在表 2-4 中。

表 2-4　PNN 识别器诊断异常模式的 ARL 性能

模式类型	向上趋势模式	向下趋势模式	向上平移模式	向下平移模式	周期循环模式
ARL	20.62	20.34	20.11	19.78	16.56
ARL 的标准差	3.27	3.30	3.84	2.82	7.29

从试验结果来看，PNN 模式识别器对于五类异常模式进行识别的 ARL 小于 25，可以认为 PNN 识别器对于过程异常的识别性能好于美国汽车工业行动 SPC 手册中规定使用的 25 个数据点的 SPC 控制图。

通过使用概率神经网络 PNN 对控制图模式的仿真实验，可以看到 PNN 对于六类基本控制图模式有很好的模式识别性能，可以在贝叶斯最小风险准则下达到最优解。同时，PNN 网络结构设计简单，易于工程实现和在现场应用，在样本信息充足的情况下不会因为神经网络的结构设计问题而导致分类与异常诊断结果的明显差异。仿真实验中对 PNN 采用的是离线训练的方式，但 PNN 的工作原理决定了其具有快速训练学习的能力，可以适应自动化生产和监控的实时性要求，潜在地能够以在线的方式在制造过程现场构建在线过程质量诊断系统。概率神经网络技术不要求过程满足正态分布假设和线性假设，适用于对复杂过程进行异常分析和诊断，对过程现场有更大的适应能力。

使用 PNN 对控制图异常/正常的模式识别，可以判断过程的状态，但还需要进一步准确确定异常模式的其他信息，如异常控制图模式的平移幅度、循环周期等，以便为过程改进提供更多的参考信息；过程现场存在着各种控制图基本模式的组合模式，对此也需要加以进行高效识别与诊断；本书对于控制图的模式识别率是很高的，但可以考虑进一步采取其他更能够提取控制图模式特征的方法加以提高，以便最大限度地减少进行过程质量诊断的成本。

2.5　本章小结

神经网络是人工智能技术的重要内容之一，被成功地应用于工程技术、经济管理、金融等众多领域。本章首先介绍了传统控制图的基本概念及其使用方法与流程，然后介绍了控制图模式现象及其仿真描述方法。在分析以往使用神经网络进行控制图模式识别方法的缺点的基础上,提出使用概率神经网络解决神经网络设计困难、识别率不高等问题。介绍了神经网络的结构和学习方式,重点介绍了概率神经网络的工作原理和性能特点。研究使用概率神经网络对西方电气公司提出的六种基本控制图模式进行模式识别的问题,通过仿真实验对该方法进行性能评价,实验结果表明,在过程质量诊断方面,概率神经网络模式识别技术具有独特的优势和优良的性能。概率神经网络模式识别技术为大批量生产条件下的过程异常诊断问题的解决,找到了一条简单高效的新的解决途径。

第 3 章 基于 LS-SVM 的小样本过程质量诊断

目前,日益多样化的市场需求使得越来越多的制造过程需要在多品种小批量的生产条件下进行。基于传统统计学的 SPC 过程质量诊断技术并不适用于小批量生产的制造环境。质量管理中必须寻找新的质量诊断技术,在生产样本数量有限的条件下,使用尽可能少的测试样本,对制造过程实施快速、高效诊断。

本章针对小样本条件下的制造过程质量诊断问题,提出使用最小二乘支持向量机技术来识别控制图模式并提出使用粒子群算法和遗传算法优化最小二乘支持向量机的参数。通过以上方法,解决了最小二乘支持向量机参数优化选择困难的问题,提高了使用最小二乘支持向量机对于各类控制图模式的识别率。

3.1 制造过程的小样本质量诊断问题

多品种、小批量生产是现代制造业的重要生产模式,在某些行业(如船舶制造业,航空制造业)还需要进行单件生产,这种情况下获取测试数据相当困难,甚至是不可能的。但以休哈特控制图为代表的传统 SPC 控制图建立在大数定理的基础之上,其对过程异常诊断只有在大样本条件下才具备很好的可靠性;在不具备大样本的条件下,传统的基于统计学的过程质量诊断方法存在着一定的局限性,因为在小批量生产条件下无法获得精确估计过程参数所需的足够的测试样本,使用传统SPC控制图进行过程异常诊断时发生"误发警报"的概率明显增加,直接导致生产的停机次数和调整增多,降低了生产的连续性和效率,导致企业生产成本的增加和利润的降低。同时,即使在大批量生产条件下,对过程进行质量监控和诊断时,尽可能减少测试样本的数量,也是进行质量管理的基本要求。这是因为,使用的样本数据需要通过对产品进行测试而得到,需要的样本数据越多,过程所需的测试费用成本和时间成本就越高,这是制造企业所不愿意承担的,不符合质量管理的经济性要求。因此,如何适应制造业小批量生产的特点、实现在小样本条件下对过程进行有效质量监控,同时设法减少样本测试成本和时间,是研究过程质量诊断技术必须考虑的主要问题之一。

另外,本书中提到的"小样本",可以认为是一个数理统计学中的概念,在数理统计学的研究中,一般认为使用的样本量超过30(或者50)就属于大样本,否则属于小样本。制造业中的"小批量"生产的概念,一般理解为批量小于100件的生产,代表性的行业有航空、航天工业、船舶工业等。本书在叙述有关内容时遵循并使用了上述概念。

3.1.1 小样本SPC控制图

针对如何在小样本条件下使用SPC控制图进行过程质量诊断的问题,人们提出了很多小样本SPC的方法加以解决,这些方法大致被分为三类[117]:数据变换法、过程建模法、贝叶斯方法。

1)数据变换法

数据变换法依赖于相似生产工序的信息,通过数据变换来将统计变量转化成标准正态变量,以便使用传统的SPC方法对过程进行异常诊断。Quesenberry提出的Q控制图是这种方法的代表。但在不存在相似性工序的情况下,这种方法无法使用。此外,这种方法对过程异常的诊断能力不足[118]。

2)过程建模法

过程建模的方法充分考虑了过程的历史信息(模型),对小的质量特性漂移具有较好的检测能力,但这种方法仍然需要一定的样本来估计过程的统计参数。基于过程模型的SPC技术包括CUSUM图和EWMA图。

3)贝叶斯方法

基于贝叶斯预测理论的SPC方法首先需要根据历史数据信息建立过程的动态模型,然后根据过程的先验信息计算后验分布,最后结合专家经验对过程的状态进行判断和决策。该方法强调重视使用主观的专家经验来达到在小样本条件下诊断过程的目的,但专家经验存在难以掌握和获取的问题。

上述小样本SPC方法都只是适用于特定的场合,并存在一定的技术缺陷,并没有完全解决小样本条件下的过程诊断问题。

3.1.2 支持向量机与小样本质量诊断

使用神经网络这一人工智能技术来诊断过程异常的方法虽然有很多的优点,但ANN识别器需要通过机器学习才能获得所需要的判断和分类能力,ANN识别器的训练和测试仍然需要使用大量的样本数据,因此,神经网络技术只适合在大批量生产条件下进行过程质量诊断,无法应用于小批量生产过程。另外,从神经网络技术本身的特点来讲,神经网络技术基于经验风险最小化原理进行训练,普遍具有训练时间长、训练结果泛化能力差、容易陷入局部最小化的缺点,不适用于对动态变化的过程进行异常诊断。目前,没有一种单一的万能的诊断技术可以解决多品种、小批量生产条件下的质量诊断问题,必须结合其他方法和具体的制造过程工艺特点,从系统的观点出发,对制造过程进行全面分析,采取灵活的策略进行过程异常诊断。

近年来,Vapnik提出了统计学习理论[42](Statistical Learning Theory,SLT)和基于统计学习理论的支持向量机(Support Vector Machine,SVM)技术,成为人工智能领域的研究热点。统计学习理论是一种小样本统计理论,具有坚实的理论基础,着重解决在小样本情况下的机器学习的问题。支持向量机是在统计学习理论基础之上发展起来的学习算法。SVM 基于结构风险最小化原则训练样本误差,可以很好地解决小样本、非线性和高维模式识别等实际问题,并且具有较好的泛化能力。在过程质量诊断领域,可以很好地应用于小样本质量诊断问题,有利于在该领域的推广应用。

3.2 有限样本条件下的统计学习理论

3.2.1 机器学习问题概述

统计学习理论[42,119]是 20 世纪 60 年代后期发展起来的新一代机器学习方法,可以在有限样本条件下达到最优解。机器学习就是从给定的观测数据(训练样本)出发寻找系统输入输出之间的规律,使得机器学习机可以利用这些规律对未来数据(或者无法观测的数据)做出尽可能准确的预测。通过机器学习需要解决三类基本问题:模式识别、函数逼近和概率密度估计。

机器学习的方法有多种,比如,决策树、群智能算法、神经网络等。机器学习并没有形成统一的理论框架,目前,机器学习的方法被大致分为三类:经典的参数统计估计方法,经验的非线性方法,统计学习理论。

1)经典的参数统计估计方法

其局限在于需要已知样本的分布形式,实际应用中难以真正做到,而且建立在样本趋于无穷大时的渐进理论之上,而实际应用中样本的数目往往是有限的。

2) 经验的非线性方法(如人工神经网络)

该方法利用已知样本建立非线性模型,克服了传统参数估计的困难,但缺乏统一的数学理论基础,使用该方法得到的结论往往难以解释,并且在实际应用中存在过学习问题,难以保证模型的泛化能力。

3) 统计学习理论

SLT 理论是为了解决有限样本学习问题而提出的一套系统的理论,可以很好地解决神经网络理论中存在的结构选择困难、局部最小点、训练结果泛化能力差等问题。在 SLT 理论上发展起来的支持向量机技术在理论和应用中表现出了优于已有方法的性能,正在成为机器学习领域里新的研究热点。

机器学习问题的基本模型如图 3-1 所示。

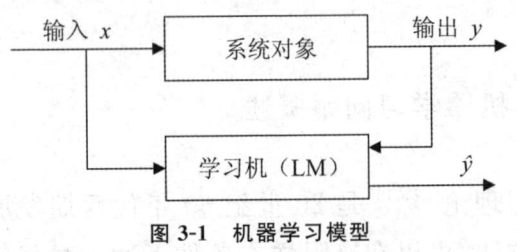

图 3-1 机器学习模型

设系统的输入 x 和输出 y 之间存在某种未知的关系,这种关系可以用某个未知联合概率分布 $P(x,y)$ 来表示。机器学习就是通过 n 个独立同分布的观测样本 $\{(x_1,y_1),(x_2,y_2),\cdots,(x_n,y_n)\}$,在一组函数 $\{f(x,\omega)\}$ 中按照某个准则确定一个最优的函数 $f(x,\omega_0)$,作为对系统输入输出关系的一种估计。其中,$f(x,\omega)$ 被称为学习机,确定学习机的参数 ω 的过程称为"训练"。

3.2.2 经验风险最小化机器学习的问题

3.2.2.1 期望风险和经验风险

由于函数 $f(x,\omega)$ 只是系统输入输出关系的一种估计,使用

$f(x,\omega)$ 对系统的输出进行预测会造成一定的损失,这种损失可以使用损失函数来度量,损失函数使用 $L(y,f(x,\omega))$ 来表示。一个学习机 $f(x,\omega)$ 的损失函数的期望被称为期望风险 $R(\omega)$（或真实风险）,

$$R(\omega) = \int L(y,f(x,\omega))\mathrm{d}P(x,y) \tag{3.1}$$

机器学习问题中学习的目的是使得期望风险 $R(\omega)$ 越小越好,但计算期望风险 $R(\omega)$ 需要使用联合概率密度 $P(x,y)$,实际问题中 $P(x,y)$ 多数情况下属于未知,期望风险 $R(\omega)$ 无法通过计算来获得。因此,人们使用考虑经验风险 $R_{\mathrm{emp}}(\omega)$ 来代替期望风险 $R(\omega)$ 进行研究,学习的准则变为"经验风险最小化"（Empirical Risk Minimization,ERM）。经验风险 $R_{\mathrm{emp}}(\omega)$ 被定义为训练样本集上的平均误差,即

$$R_{\mathrm{emp}}(\omega) = \frac{1}{n}\sum_{i=1}^{n}L(y_i,f(x_i,\omega)) \tag{3.2}$$

3.2.2.2 基于经验风险最小化学习的缺陷

传统的学习方法（如,神经网络）长期采用了所谓经验风险最小化（ERM）准则作为研究的起点,但这一做法只是直观上显得合理,并没有经过充分的理论论证。许多研究基于 ERM 准则,只是关注于使得训练误差最小,但不久发现这一做法存在一定的问题,主要包括下面几个方面。

（1）经验风险最小不等于期望风险最小,因此根据经验风险最小化训练出来的学习机（如,神经网络）不能保证其泛化能力。

（2）根据大数定理,经验风险只有在样本数目为无穷大时才趋于期望风险,学习机的训练只有在大样本的情况下才能保证其训练的结果具有一定的泛化性。

（3）当经验风险过小时,学习机的泛化能力常常反而下降,存在所谓"过学习,过适应"现象。

可见,基于经验风险最小化的学习原则并不能保证在小样本情况下学习机的学习训练效果和学习的泛化性,需要采用其他的

机器学习理论来加以解决。

3.2.3 结构风险最小化的机器学习方法

3.2.3.1 VC维和泛化能力的界

统计学习理论就是在小样本条件下机器学习的有效理论,该理论的成果包括经验风险最小化原则成立的条件、有限样本条件下经验风险与期望风险的关系以及如何利用这些理论找到新的学习原则和方法等。其中,"泛化能力的界"和"VC维"(Vapnik Chervonenkis Dimension)是其重要的理论成果。

VC维是由学习机的分类函数实现的分类能力的测度,即分类函数集$\{f(x,\omega)\}$的VC维是能被$\{f(x,\omega)\}$正确分类的训练样本的最大数量。VC维越高的学习机就越复杂。目前尚没有用于计算任意函数集VC维的统一的方法,只对一些特殊函数集合知道其VC维。比如,n维实数空间中线性分类器的VC维是$n+1$。

统计学习理论中的"泛化能力的界"的概念,指的是经验风险和期望风险之间的关系。设学习机的VC维是h,训练样本的数目为n,损失函数的取值概率为$1-\eta(0 \leqslant \eta \leqslant 1)$,对于两类分类问题,经验风险和期望风险之间满足如下关系,

$$R(\omega) \leqslant R_{\text{emp}}(\omega) + \sqrt{\frac{h(\log(2n/h)+1)-\log(\eta/4)}{n}} \quad (3.3)$$

或者简单写为

$$R(\omega) \leqslant R_{\text{emp}}(\omega) + \phi(n/h) \quad (3.4)$$

上述两式的右端被称为"风险界,泛化能力的界",其中的第一项是由训练误差构成的经验风险,第二项称为置信范围,是随n/h增大而减小的函数。不难看出,对于有限训练样本的情况,学习机的VC维h越高,学习机越复杂,则置信范围$\phi(n/h)$越大,导致了期望风险$R(\omega)$和经验风险$R_{\text{emp}}(\omega)$的差别越大,这就是神

经网络过学习问题的原因。良好的机器学习方法不但要使得经验风险尽量减小,而且应当同时尽量减小 VC 维,以便减小置信范围 $\phi(n/h)$,才能使得实际风险 $R_{emp}(\omega)$ 尽量减小,最终使得学习机具有良好的泛化能力。

通常人们使用神经网络构建学习机时,总是先利用个人经验确定学习机的结构,这相当于首先选定了分类函数集 $\{f(x,\omega)\}$ 的一个子集,相当于确定了置信范围 $\phi(n/h)$;然后通过训练确定神经网络的权值,相当于将经验风险 $R_{emp}(\omega)$ 进行最小化处理。但神经网络的结构的设计具有一定盲目性,无法确定泛化的置信范围 $\phi(n/h)$,导致即使经验风险很小(或者最小),但可能学习机的泛化能力很差。

3.2.3.2 结构风险最小化学习原理

由于基于经验风险最小化训练的学习机不能保证学习机的泛化能力,人们通过研究发现按照结构风险最小化(Structural Risk Minimization,SRM)的原理寻找学习机是更合理的选择。基于结构风险最小化的机器学习方法就是,将分类函数集 $S = \{f(x,\omega)\}$ 分解成为嵌套函数子集序列:

$$S_1 \subset S_2 \subset \cdots \subset S_k \subset \cdots \subset S \tag{3.5}$$

并且要求各个子集按照它们的 VC 维的大小(即 $\phi(l/h)$ 的大小)排列:

$$h_1 \leqslant h_2 \leqslant \cdots \leqslant S_k \leqslant \cdots \tag{3.6}$$

这样处理的结果,使得在同一个子集中,置信范围就相同。结构风险最小化机器学习的思路就是在每个子集中寻找最小经验风险,在各子集之间折中考虑经验风险和置信界限,取得实际(期望)风险最小。为了使机器学习的期望风险最小,只需要首先选择经验风险与置信范围之和最小的子集,然后在这个子集中,使得期望风险最小的函数就是要求的最优函数。支持向量机技术就是统计学习理论的一种具体实现算法。

3.3 支持向量机理论

SVM 技术[120-123]是统计学习理论中的最新成果,是一种基于结构风险最小化的学习算法,其在有限样本条件下的出色学习能力具有很大的实用性。该算法属于二次优化问题,因此获得了全局最优解。SVM 理论在 20 世纪 90 年代由 Vapnik 等人提出,是人工智能和机器学习领域的研究热点,目前仍在不断发展,在各领域的应用成果越来越多。

3.3.1 最优分类超平面和支持向量机

可以从两类数据的线性分类问题来理解支持向量机的基本思想和特点。设 $f(x) = \omega^T x + b = 0$ 是一个分类决策超平面 H,ω 是其法矢量,决定了超平面的方向,b 决定超平面的位置。当两类问题训练样本集为 $S = (x_1, y_1), (x_2, y_2), \cdots, (x_n, y_n)$,其中,$x_i \in R^d, y_i = \{1, -1\}, i = 1, 2, \cdots, n$。训练样本的线性可分是指存在参数对 (ω, b),使得

$$y_i = \text{sgn}(\omega^T x_i + b) \quad (i = 1, 2, \cdots, n) \tag{3.7}$$

两类样本线性可分的情况可以用图 3-2 来表示。

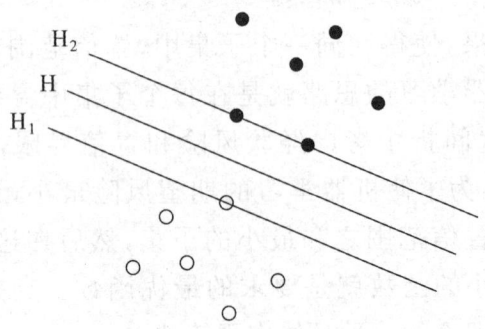

图 3-2 SVM 最优分类面原理

图 3-2 中的白点和黑点分别代表两类样本数据，H 为分类决策超平面，H_1 和 H_2 分别表示通过各类样本中距离 H 最近的样本并且平行于 H 的面。H_1 和 H_2 的距离称为分类间隔（Margin）。直观上看，显然当 H 位于 H_1 和 H_2 的中间时，分类效果最好，最优分类面就是将两类样本正确分开且使得 H_1 和 H_2 之间的间隔最大的那个分类面。由于 H_1 和 H_2 之间的距离为：

$$D = \frac{2}{\|\omega\|} \tag{3.8}$$

则求取最优分类超平面的问题转化为如下优化问题：

$$\begin{cases} \max \quad \dfrac{2}{\|\omega\|} \\ \text{s.t.} \quad y_i(\omega^T x_i + b) \geqslant 1 \quad (i=1,2,\cdots,n) \end{cases} \tag{3.9}$$

或者等同于：

$$\begin{cases} \min \quad \Phi(\omega) = \dfrac{1}{2}\omega^T \omega \\ \text{s.t.} \quad y_i(\omega^T x_i + b) \geqslant 1 \quad (i=1,2,\cdots,n) \end{cases} \tag{3.10}$$

构造拉格朗日多项式：

$$L(\omega,b,\alpha) = \frac{1}{2}\omega^T \omega - \sum_{i=1}^{n} \alpha_i(y_i(\omega^T x_i + b) - 1) \tag{3.11}$$

其中，$\alpha_i \geqslant 0$，是非负的拉格朗日乘子。根据拉格朗日方法和对偶理论，得到上面优化问题的对偶问题：

$$\begin{cases} \max \quad Q(\alpha) = \sum_{i=1}^{n} \alpha_i - \dfrac{1}{2}\sum_{i,j=1}^{n} \alpha_i \alpha_j y_i y_j x_i^T x_i \\ \text{s.t.} \quad \sum_{i=1}^{n} y_i \alpha_i = 0 \\ \alpha_i \geqslant 0, i=1,2,\cdots,n \end{cases} \tag{3.12}$$

求取上述凸二次规划问题的全局最优解可得：

$$\omega = \sum_{i=1}^{n} y_i \alpha'_i x_i \tag{3.13}$$

同时上述优化问题的全局最优解 α'_i, ω', b' 必须满足 KKT（Karush-Kuhn-Tucker）条件，即：

$$\alpha'_i(y_i(\omega'^T x_i + b') - 1) = 0 \quad (i = 1, 2, \cdots, n) \quad (3.14)$$

任取 $\alpha'_i \neq 0$，可求得 b'。可以证明，根据式(3.13)可使得大部分 α_i 为零，其解具有稀疏性。我们将 α_i 不为零所对应的训练样本称为"支持向量"（在图 3-2 中，支持向量位于 H_1 和 H_2 上）。此时，权重向量中只包含支持向量。

$$\omega = \sum_{i \in SV} y_i \alpha'_i x_i \quad (3.15)$$

最后，优化超平面可以表示为

$$g(x) = \omega^T x + b' = \sum_{i \in SV} y_i \alpha'_i x^T x + b' \quad (3.16)$$

式(3.16)就是求得的分类函数表达式。可见，线性可分支持向量机具有的特点如下：

（1）分类超平面仅由少数支持向量决定。

（2）最优分类面的数据出现在内积中，这使得使用该技术在特征空间中找到并使用超平面成为可能。

（3）Vapnik 证明了线性分类器的 VC 维满足：

$$h \leqslant \|\omega\|^2 r^2 + 1 \quad (3.17)$$

其中，r 为包络训练样本的最小球半径。这表明 SVM 在训练样本正确被划分的同时，又最小化了分类平面的 VC 维，属于结构风险最小化的训练方式，训练结果具有很强的泛化性。

最大间隔线性分类器是构造更加复杂的支持向量机的起点。对于训练样本线性不可分的学习问题，SVM 技术对 n 个训练点 $((x_i, y_i), i = 1, 2, \cdots, n)$ 引入松弛变量 $\xi_i \geqslant 0 (i = 1, 2, \cdots, n)$，对目标函数引入惩罚因子 C，使得求取最优分类面的优化问题变为：

$$\begin{cases} \min \quad \Phi(\omega) = \frac{1}{2} \omega^T \omega + C \sum_{i=1}^{n} \xi_i \\ \text{s.t.} \quad y_i(\omega^T x_i + b) \geqslant 1 - \xi_i \\ \xi_i \geqslant 0, i = 1, \cdots, n \end{cases} \quad (3.18)$$

式中，ξ_i 为错分随机误差，$\sum_{i=1}^{n} \xi_i$ 表示训练样本集被错分的情况，

惩罚因子 C 控制对错分样本惩罚的程度,它折中经验误差和模型的复杂程度以使所求 SVM 具有较好的泛化能力,且当 γ 增大时,经验风险减小。对于训练样本线性不可分的情况,其最优分类超平面的形式同线性可分的情况是一样的,但支持向量机的约束条件 $\alpha_i \geqslant 0$ 被替换为更强的 $0 \leqslant \alpha_i \leqslant C$。

3.3.2 支持向量机及其结构

对于非线性分类问题,可以通过非线性变换 $\phi(\cdot)$ 将样本映射到某个高维特征空间 F 中($\phi:x \to \phi(x), \phi:R^d \to F$),成为高维空间的线性分类问题,并在 F 中构造最优分类面函数,

$$y_i = \omega^T \phi(x_i) + b \tag{3.19}$$

但是这种映射变换可能比较复杂。注意到,SVM 的算法中,最优分类超平面的寻优函数和分类函数中只是涉及了内积运算 $x_i \cdot x_j$,这样,在高维空间实际上只需进行内积运算 $\phi(x_i) \cdot \phi(x_j)$,而这种内积运算是可以用原输入空间中的函数实现的。根据泛函理论,只要核函数 $k(x_i, x_j)$ 满足 Mercer 条件,它就对应某一变换空间中的内积。可见,采用适当的核函数,SVM 就可以通过非线性变换将输入空间变换到高维空间,然后在新的高位空间中求解最优分类面,线性可分情况下的内积运算变为 $k(x_i, x_j) = \phi(x_i) \cdot \phi(x_j)$,此时支持向量机分类函数变为:

$$y = g(x) = \text{sgn}(\sum_{i \in SV} y_i \alpha'_i k(x_i, x) + b') \tag{3.20}$$

由上式可知,支持向量机分类函数在形式上类似于神经网络,该神经网络的中间节点分别对应于一个支持向量,支持向量机的结构如图 3-3 所示。

图 3-3 中,x 表示 SVM 的输入向量,$x = (x^1, x^2, \cdots, x^d)$,$d$ 为输入向量的维数,s 为 SVM 的支持矢量的个数。使用不同的核函数就会形成不同类型的 SVM,目前经常使用的核函数包括以下几种。

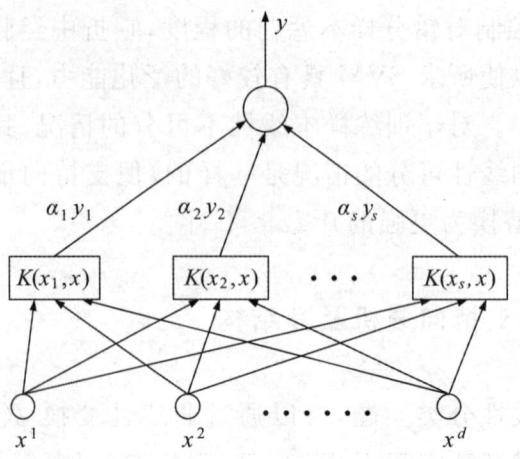

图 3-3 支持向量机结构示意图

1）多项式核函数
$$K(x, x_i) = [(x \cdot x_i) + 1]^q \qquad (3.21)$$
此时得到的 SVM 就是一个 q 阶多项式分类器。

2）径向基核函数
$$K(x, x_i) = \exp\left\{-\frac{\|x - x_i\|^2}{2\sigma^2}\right\} \qquad (3.22)$$
则 SVM 就是一个径向基分类器，每一个基函数的中心对应一个支持向量。

3）S 形函数核函数
$$K(x, x_i) = \tanh(\upsilon(x \cdot x_i) + c) \qquad (3.23)$$
使用 S 形函数核函数的 SVM 相当于一个自动生成的、两层的多层感知器 NN。

显然，不同的核函数对于 SVM 的性能影响是不同的，在实际应用中核函数对于问题的解决效果的影响主要通过对比研究来确定。

3.3.3 最小二乘支持向量机

3.3.3.1 LS-SVM 原理

支持向量机有许多发展与变化的形式，其中较为常见的是最

第 3 章　基于 LS-SVM 的小样本过程质量诊断

小二乘支持向量机(LS-SVM)。相对于其他类型的 SVM，LS-SVM 在求解的速度方面具有一定的优势，但该方法当训练样本增多时，支持向量解的稀疏性会变差。

设训练样本数据集为 $\{x_i,y_i\}_{i=1}^n$，n 为样本容量，输入矢量数据 $x_i \in R^d$，其对应的输出类别标量数据 $y_i \in R$，表示输入数据所对应的类别。Suykens 设计的 LS-SVM 使用训练集错分误差的二范数 $\sum_i^N \xi_i^2$ 代替标准 SVM 中的错分误差之和 $\sum_i^N \xi_i$，形成下面的优化问题：

$$\begin{cases} \min \quad J(\omega,\xi) = \frac{1}{2}\omega^T\omega + \frac{1}{2}\gamma\sum_{i=1}^N \xi^2 \\ \text{s.t.} \quad y_i = \omega^T\phi(x_i) + b + \xi_i, i=1,2,\cdots,n \end{cases} \quad (3.24)$$

式中，ξ_i 为错分随机误差，$\gamma>0$ 为对错分误差平方和的惩罚系数。构造拉格朗日函数，

$$L(\omega,\xi,\alpha,b) = J(\omega,\xi) - \sum_{i=1}^n \alpha_i(\omega^T\phi(x_i) + b + \xi_i - y_i)$$

$$(3.25)$$

其中，$\alpha = [\alpha_1,\alpha_2,\cdots,\alpha_n]^T, i=1,2,\cdots,n, \alpha$ 为拉格朗日乘子(Suykens 称其为支持值，在 Karush-Kuhn-Tucker 等式约束条件下，其值可正可负)，根据 KKT 条件，对式(3.25)分别对 ω,ξ,α,b 求偏导并置零，得到下面的优化条件：

$$\begin{cases} \omega = \sum_{i=1}^n \alpha_i\phi(x_i) \\ \alpha_i = \gamma\xi_i, i=1,2,\cdots,n \\ y_i = \omega^T\phi(x_i) + b + \xi_i, i=1,2,\cdots,n \\ \sum_{i=1}^N \alpha_i = 0 \end{cases} \quad (3.26)$$

消去 ω,ξ 后，优化问题变为求解如下线性方程：

$$\begin{pmatrix} \Omega + \gamma^{-1}I & l_n \\ l_n^T & 0 \end{pmatrix} \begin{pmatrix} \alpha \\ b \end{pmatrix} = \begin{pmatrix} y \\ 0 \end{pmatrix} \quad (3.27)$$

其中，$l_n = [1,1,\cdots,1]^T$，$\alpha = [\alpha_1,\alpha_2,\cdots,\alpha_n]^T$，$y = [y_1,y_2,\cdots,y_n]^T$，$\Omega$ 为一个方阵，其元素 $\Omega_{ij} = K(x_i,y_j)$，$i,j = 1,2,\cdots,n$，$K(\cdot)$ 定义为满足 Mercer 条件的核函数。解方程(3.27)求出 α,b 后，可得如下线性回归模型：

$$y(x) = \sum_{i=1}^{n}\alpha_i K(x,x_i) + b \tag{3.28}$$

本书在下面的研究中，将 $K(\cdot)$ 选取为 RBF 核函数 $K(x,y) = \exp\left\{-\dfrac{\|x-y\|^2}{2\sigma^2}\right\}$。LS-SVM 的参数 γ 和 σ 对其应用于模式识别时的性能会产生明显影响，应当使用适当的方法获得。

3.3.3.2 LS-SVM 的加权改进

在进行控制图模式识别时，来自于过程现场的训练数据中常常包含异常值，或者过程数据不能够严格满足正态分布的假设。由于 LS-SVM 采用 $\sum\limits_{i}^{N}\xi_i^2$ 作为评价函数，当训练数据中存在异常值或非高斯噪声时，求得的分类函数可能失去稳健性。针对这一问题，可以采用加权的方法使最小二乘向量机[128]来获得解的稳健性。该方法在式(3.26)中的松弛误差变量 $\alpha_i = \gamma\xi_i$ 引入加权因子 v_i，可得下面的最优分类超平面问题：

$$\begin{cases} \min \quad J(\omega^*,\xi^*) = \dfrac{1}{2}\omega^{*T}\omega^* + \dfrac{1}{2}\gamma\sum_{i=1}^{n}v_i\xi^2 \\ \text{s. t.} \quad y_i = \omega^{*T}\phi(x_i) + b^* + \xi_i^* \end{cases} \tag{3.29}$$

其相应的拉格朗日函数为：

$$L(\omega^*,\xi^*,\alpha^*,b^*) = J(\omega^*,\xi^*) - \sum_{i=1}^{n}\alpha_i^*(\omega^{*T}\phi(x_i) + b^* + \xi_i^* - y_i) \tag{3.30}$$

使用"*"表示待定未知变量。根据 KKT 条件并消去 ω^*，ξ^*，可得如下线性方程：

$$\begin{bmatrix} 0 & l_n^T \\ l_N & \Omega + V_\gamma \end{bmatrix}\begin{bmatrix} b^* \\ \alpha^* \end{bmatrix} = \begin{bmatrix} 0 \\ y \end{bmatrix} \tag{3.31}$$

式中，$V_\gamma = \text{diag}[1/v_1, 1/v_2, \cdots, 1/v_n]$，加权因子 v_i 的取值同误差 ξ_i 相关，可以使用下面的式子得到：

$$v_i = \begin{cases} 1, & \text{当 } |\xi_i/\hat{s}| \leqslant c_1 \\ \dfrac{c_2 - |\xi_i/\hat{s}|}{c_2 - c_1}, & \text{当 } c_1 \leqslant |\xi_i/\hat{s}| \leqslant c_2 \\ 10^{-4}, & \text{其他} \end{cases} \quad (3.32)$$

其中，ξ_i 可以由式(3.26)求出。c_1, c_2 为常数，它们的典型经验取值为 $c_1 = 2.5, c_2 = 3$；$\hat{s}$ 是误差变量 ξ_i 的稳健估计值[129]，

$$\hat{s} = 1.483 MAD(\xi_i) \quad (3.33)$$

MAD 表示中值绝对偏差。通过使用稳健统计学的方法对上述误差项 ξ_i 的处理，使得当 ξ_i 中存在异常值或不满足高斯特性时，获得了 LS-SVM 模型参数的稳健估计值。由式(3.31)求得 b^* 和 α_i^* 后，且令

$$f(x) = \sum_{i=1}^{n} y_i \alpha_i^* K(x, x_i) + b^* \quad (3.34)$$

$f(x)$ 是由核 $K(x,z)$ 隐式定义的特征空间中的最大间隔超平面，决策规则由符号函数 $\text{sgn}[f(x)]$ 给出，$\text{sgn}(x) = \begin{cases} 1, x > 0 \\ -1, x < 0 \end{cases}$。当使用 RBF 核函数，$b^*$ 已经包含在核函数 $K(x, x_i)$ 中[54]，此时，SVM 的分类决策函数为：

$$f(x) = \sum_{i=1}^{n} y_i \alpha_i^* K(x, x_i) \quad (3.35)$$

3.4 基于 LS-SVM 控制图模式识别的过程异常诊断

人工神经网络的智能过程异常诊断方法在处理过程的复杂特性方面具有较大的优势，属于故障诊断领域的流行方法，并且取得了相当的研究成果，然而神经网络过程异常诊断方法的理论基础从本质上讲的是传统的统计学，其统计规律只有当训练样本

接近无限大时才具有较好的可靠性。然而,在处理实际的过程质量诊断问题时,由于小批量生产和测试成本等条件的限制,只能得到非常有限的过程故障样本数据,从而导致神经网络质量诊断中容易出现过多误诊和漏诊,影响过程质量诊断的效果。同时人工神经网络本身在算法上存在过学习、泛化能力差、训练速度慢等缺点。

支持向量机(SVM)采用结构风险最小化机器学习原理,同时考虑训练误差和学习机的泛化能力,在解决小样本及非线性问题上具有独特的优势,特别适合建立过程诊断模型。本节的研究内容采用 LS-SVM 技术来设计 LS-SVM 模式识别器,对六类控制图模式进行模式识别,有效实现在有限样本条件下对制造过程进行过程异常诊断。

3.4.1 多分类 LS-SVM 模式识别器的构建

传统的 SVM(包括 LS-SVM)是基于解决两类数据的分类问题的,而在本书的研究中,对六类控制图的模式进行识别,需要使用具有多类分类能力的支持向量机分类器。从解决问题的思路上讲,构建一个多分类 SVM 分类器的方法大致分为两类,要么需组合多个两类 SVM 分类器,要么需构造一个更大的最优化问题来形成多分类的 SVM 分类器,由此产生了很多的具体的构造多分类 SVM 分类器的方法,每种多类 SVM 已应用到实际问题中并取得了不错的效果,但每种方法也都或多或少存在着缺陷。构建多类 SVM 分类器的常见的方法有两种,"一对多"(One Against All, OAA)算法和"一对一"(One Against One, OAO)算法[130-133]。通过这两种算法,扩展了传统的两类 SVM 分类器的分类能力。

3.4.1.1 "一对多"算法

"一对多"算法构建多类 SVM 分类器的思路简单,对于

$K(K>2)$ 类分类问题,该算法对每类数据建立一个 SVM 分类器,这个分类器可以将该类数据同其他数据区分开来,为了将所有类别的数据区分开,总共需要建立 K 个分类器。

设将第 $i(i\leqslant K)$ 类数据同其他 $K-1$ 类数据区分开来的 SVM 分类器的分类决策函数用 $(\omega_i^T j\phi(x)+b_i)$ 表示,并且将第 i 个 SVM 分类器中第 i 类的训练样本作为正的训练样本,而将其他的样本作为负的训练样本。为了判断一个质量控制图测试样本 x_t 属于哪一类,需要将该测试样本 x_t 依次输入到这 K 个分类器中,令 $f_i(x_t)=\text{sgn}(\omega_i^T\phi(x_t)+b_i)$,最后,样本 x_t 的类别被判断为下面的类别:$\underset{i=1,\cdots,K}{\text{argmax}} f_i(x_t)$。

从"一对多"SVM 多类器的设计过程看,该算法简单有效,SVM 的训练时间较短,当数据类别较多时具有优势。但其缺点在于:①当类别数较多时,必然是某一类的训练样本将大大少于其他各类训练样本的总和,这种训练样本间的不均衡将对最后的分类精度产生不良影响;②分类过程中会出现较多的"误分、拒分"的测试样本;③算法的泛化能力较"一对一"多类 SVM 分类器差。

3.4.1.2 "一对一"算法

使用该算法在构建多类 SVM 分类器时,对于 $K(K>2)$ 类分类问题,该方案首先在每两个类别的数据之间训练出一个 SVM 分类器,总共需要建立 $C_K^2=K(K-1)/2$ 个两类 SVM 分类器。对于第 i 类和第 j 类数据,其分类决策函数由 $(\omega_{ij}^T\phi(x)+b_{ij})$ 来表示。然后,对于一个质量控制图测试样本 x_t,该方法依次将 x_t 输入到这 $K(K-1)/2$ 个两类 SVM 分类器,并且采用"多者为胜"的投票法进行投票和决策。当使用对第 i 类和第 j 类两类 SVM 分类器 $(i\neq j,i,j\leqslant K)$ 判决时,如果该分类器 $\text{sgn}(\omega_{ij}^T\phi(x_t)+b_{ij})$ 判决 x_t 属于第 i 类,就给第 i 类得票数加 1,否则就给第 j 类得票加 1。$K(K-1)/2$ 个两类 SVM 分类器都对该样本进行测试,最后样本 x_t 属于得票数最多的那一类。即,对于

分类器 $f_{ij}(x) = \omega_{ij}^T \phi(x) + b_{ij}$，令 $f_i(x_t) = \sum_{j \neq i, j=1}^{K} \text{sgn}(f_{ij}(x_t))$，则测试控制图样本 x_t 最后被分到下面的类别中：

$$\underset{i=1,\cdots,K}{\arg\max} f_i(x_t)$$

"一对一"算法的优点是：由于建立每个两类 SVM 分类器只需要考虑两类样本，因此每个两分类 SVM 的训练都较为容易，其决策方法较"一对多"算法简单；另外，虽然它的决策复杂度按照类数的平方增长，但当样本数据的类别数目较少时，其分类的速度并不比传统的"一对多"方法慢，其分类精度也较"一对多"高。缺点是：①如果单个的两类 SVM 分类器不规范化，没有获得相同的并且较高的分类能力，则整个 K 类 SVM 分类器将趋向于过学习，导致推广误差可以是无限大；②两类 SVM 分类器的数目会随着类别的数目的增加而急剧增加，导致整个分类决策过程速度很慢；③同样存在"误分、拒分"的决策区域。

从训练速度上考虑，"一对一"和"一对多"两种算法，要训练的 SVM 分类器的个数之比为 $(n-1)/2$，因此当数据的类别数目较大时，采用"一对一"算法训练，需要训练的两类 SVM 分类机的数目较多，训练速度慢，因此若要采用"一对一"算法，训练数据的类别的数目不能过大。综合比较"一对多"和"一对一"多类模式识别方案，一般认为在样本类别数目不是很大时，"一对一"方案优于"一对多"方案；另外，当样本的类别数目较小时，两种算法的模式识别率几乎相同。

综合考虑"一对一"和"一对多"两种算法的优缺点，本书使用 LS-SVM 对六类控制图进行模式识别时，采用"一对一"算法建立 SPC 控制图的六类模式识别器，这样，需要首先建立 6×(6−1)/2＝15 个两类模式分类器，然后将它们组合成六类 LS-SVM 模式识别器。

3.4.2 基于 LS-SVM 的控制图模式识别仿真实验

本书通过 MATLAB 仿真实验进行 LS-SVM 的控制图模式

识别的研究并评估其性能,实验的具体情况如下。

3.4.2.1 实验过程

本书使用 LS-SVM 识别器进行六类控制图模式识别的仿真试验步骤如图 3-4 所示。对该实验做如下说明:

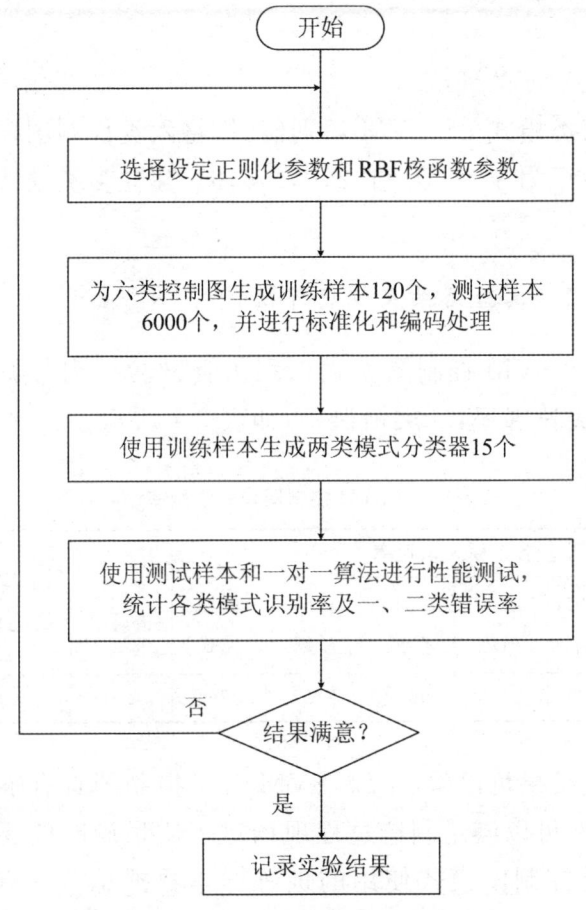

图 3-4　LS-SVM 控制图模式识别过程图

1)实验参数

使用人工交叉验证的方法反复选择 LS-SVM 的正规化参数 γ 和核函数参数 σ 的取值,最后得到 $\gamma=10, \sigma=2.2$,可以取得良好的控制图模式识别率。

2）实验数据

实验需要的训练数据和测试数据使用与第 2 章 2.4.2 节完全相同的仿真生成方法。对于六类控制图模式,每类控制图各生成 20 张控制图训练数据,共生成 $6\times20=120$ 个训练样本,并进行数据预处理。实验需要的测试数据,为每类控制图各生成 1000 张控制图训练数据,共生成 $6\times1000=6000$ 个测试样本,并进行数据预处理。

3）LS-SVM 的训练

训练两类模式 LS-SVM 识别器时,将六类控制图模式对应的输出分别标记为 $y_i=i, i=1,2,3,4,5,6$。建立多类模式识别器使用的是"一对一"算法。

3.4.2.2 实验结果及性能评估

本书 LS-SVM 控制图模式识别仿真试验得到各类控制图的识别率,一类错误率,二类错误率,如表 3-1 所示。

表 3-1 LS-SVM 控制图模式识别率(%)

各个模式识别率						一类错误率	二类错误率	总识别率
正常模式	向上趋势	向下趋势	向上阶跃	向下阶跃	周期模式			
97.2	99.0	99.5	96.2	93.3	97.4	2.0	0.88	97.1

从试验结果可以看出,LS-SVM 对于控制图具有很高的模式识别率,基本可以满足制造过程现场对于质量诊断的要求。仿真实验中六类控制图模式使用的训练样本数据总共只有 $6\times20=120$ 个,使用的训练样本的数量比通常的神经网络机器学习方法使用的数千甚至上万个训练数据大大减少,并且机器的学习速度快。实验过程中,参数 γ 和 σ 仅仅使用了人工交叉验证的方法进行选取,费时较多。如果采用进化优化算法(如遗传算法)对参数 γ 和 σ 进行优化选取,相信可以简化参数的选取过程,并取得更为理想的识别效果。

SPC 控制图模式识别是质量诊断与改进的重要方法之一,多品种小批量等先进制造过程要求质量诊断必须在有限测试样本的条件下进行。针对以上质量诊断的要求,本书提出的基于 LS-SVM 技术进行质量控制图模式识别。通过理论分析和仿真实验,可以认为,本书提出的基于 LS-SVM 控制图模式识别的质量异常诊断方法,可以避免神经网络模式识别方法基于经验风险最小化训练、存在过学习等缺点,SVM 技术在有限训练样本情况下具有的良好的泛化能力,并且可以应用于非高斯非线性制造过程。仿真试验结果表明,该方法在有限训练样本的情况下表现出了模式识别率高、训练算法学习速度快等优点。该技术的以上特性,对于降低质量诊断成本、提高质量诊断速度,以及该技术在过程现场的实际推广应用具有现实意义。

3.5 基于智能进化算法和 LS-SVM 的过程异常诊断技术

支持向量机技术基于统计学习理论,在解决小样本分类问题时,具有坚实的理论基础,保证了机器学习结果的泛化能力。然而,同其他机器学习算法一样,使用 SVM 解决实际问题的效果依赖于 SVM 学习机选择的参数值。迄今为止,人们针对这一问题进行了研究,找到了一些 SVM 学习机参数选择的方法,但这些方法多数基于个人经验和特定的知识,缺乏统一的理论指导,并没有一种方法是公认的选择 SVM 参数的最优方法。SVM 学习机的训练效果和泛化能力依赖于其参数选择的结果。如何采用一定的算法合理确定 SVM 学习机的最优参数,一直是使用支持向量机解决实际问题时需要研究解决的主要问题之一。

3.5.1 支持向量机的参数选择问题与群智能算法

3.5.1.1 SVM的参数选择问题

SVM技术中的自由参数选择与优化问题也被称为"模型选择问题"。对于标准的SVM而言，其经常使用的核函数是径向基核函数(RBF核函数，或称高斯核函数)，它需要选择和确定的参数包括核函数的宽度σ和惩罚因子C。对于最小二乘SVM，需要调整和确定的参数包括核函数的宽度σ和正则化参数γ，这两个参数必须组合成为一个整体进行优化选择和使用，但是它们对于最小二乘支持向量机性能的影响在理论上没有必然的联系，因此，在实际应用中解决这两个参数的优化选取问题具有更大的难度。

SVM参数选择方法中典型地包括"交叉验证法"(Cross Validation)[134,135]和"留一法"(Leave One Out)[136]。留一法可以看成是交叉验证法的特例。由于交叉验证技术主要依靠实验手段和个人对于训练数据分组的经验来调整参数，缺乏坚实的理论基础，参数寻优的计算量较大，难以成为理想的参数选择手段。此外，人们寻找不同的算法来优化SVM(或者LS-SVM)的参数，比如"三步搜索法"[137]，但该方法虽然可以减少搜索的步骤，但并不能保证得到的是最优解。目前，多数的支持向量机技术所采用的参数选择方法都带有较强的经验色彩，大多是使用"试凑法"(Trail and Error)并结合平时的经验，通过反复实验并不断调整参数的取值来寻找相对较优的参数。这种方法计算量较大，费时费力，而且经验性太强，缺乏理论指导，导致参数选择的效果与实际可能的最优参数取值的效果相去甚远。因此，为SVM参数选择问题找到新的不依赖于个人经验和问题背景知识的解决方法，具有实际意义。在这方面，使用群智能及进化算法来解决SVM技术中的参数选择问题是比较理想的方案。

群智能算法及进化算法是基于自然选择/遗传等生物进化机制或者社会群体性动物(蚁群、蜂群、鱼群等)的自组织活动规律而建立起来的搜索算法。与其他的搜索方法一样,进化计算进行迭代运算,但是这类算法在搜索寻优的过程中,能够从原问题的一组初始解出发,按照一定仿生进化和搜索准则进化到另外更好的一组解,再从这组解进一步搜索和进化,直到找到问题的最优解。群智能算法及进化优化算法对各个领域的优化问题采取同样的进化优化策略,能适应不同环境的问题,算法具有鲁棒性,在大多数情况下都能得到比较满意的有效解。按照不同的进化机制与策略,人们找到了多种各具特色的群智能及进化算法,遗传算法和粒子群算法是它们当中比较具有代表性的两类进化寻优算法。

3.5.1.2 群智能算法概述

人工智能技术的一个重要的分支研究领域就是进化算法和群智能算法,目前主要包括遗传算法、模拟退火算法、蚁群算法和粒子群算法及人工免疫算法和鱼群算法等[54-63]。其中,进化算法中的遗传算法出现得较早,而作为群智能算法代表的蚁群和粒子群算法,在搜索策略上具有先进性。

遗传算法形成于20世纪70年代初期,主要由美国Michigan大学的Holland教授与其同事、学生通过模仿生物的遗传和进化机制而形成了一整套优化理论。遗传算法的特点是采用简单的编码技术来表示各种复杂问题的结构,通过对一组编码表示进行简单的遗传操作机理和优胜劣汰的自然选择机理来指导学习进化与确定下一步搜索的方向。算法从一组随机产生的初始群体开始搜索过程,以适应度函数为决策的依据,采用基于适应值比例的选择策略在当前种群中选择个体,通过杂交和变异操作来产生下一代种群。如此模仿生命的进化优化过程一代一代演化下去,直到满足设定的终止条件为止。其优点是:①与问题的领域无关,具有快速随机的搜索能力,应用领域十分广泛;②搜索从初始群体出发,具有很好的全局搜索能力;③搜索使用评价函数启发进行,约束条件缩小,不需要问题的背景知识,应用范围大大扩展;④使用概率随机搜索机制进行迭代,具有内在并行搜索机制;⑤具有可扩展性,容易与其他算法结合得到性能更加优良的新算法。其存在的主要缺点是,遗传算法容易陷入局部最优解、出现早熟现象,收敛速度较慢。模拟退火算法(Simulated Annealing Algorithm,SAA)是基于 Mente Carlo 迭代求解策略的一种随机寻优算法,其最大优点在于该算法执行过程中即便落入局部最优,理论上经过足够长的时间后也可跳过局部最优解,最终收敛到全局最优解,但是,由于模拟退火算法对整个搜索空间的状况了解不充分,具有运算效率低的缺点。此外,人工免疫算法(Artificial Immune Algorithm,AIA)模仿生物免疫系统的自适应机制和排除机体的抗原性异物机制,其选择算子模拟了自然免疫系统的抗体繁殖策略,可以克服遗传算法容易出现早熟收敛和陷入局部最优的缺点,但 AIA 的缺点也很突出,就是它的运行速度和收敛速度都较慢。

随着人们对于生命本质的不断了解,社会性动物(蚁群、蜂群、鸟群等)的自组织行为引起了人们的关注和研究兴趣。人们通过模拟自然界生物群体的智能自组织行为,研究出了新兴的所谓群智能算法。群智能理论研究领域具有代表性的两种算法是蚁群算法和粒子群优化算法。蚁群算法擅长解决离散数值的优化问题,粒子群算法擅长解决连续数值的优化问题。

蚁群算法由意大利学者 M. Dorigo 等人于 1991 年首先提出,该算法是对蚂蚁群落食物寻找/采集过程的模拟,已成功解决了许多离散优化问题。粒子群优化算法(Particles Swarm Optimization,PSO)最早是由 Kenney 与 Eberhart 于 1995 年提出的,该算法起源于对简单社会系统的模拟,最初是模拟鸟群觅食的过程,是一种性能优良的解决优化问题的工具。我国李晓磊博士于 2002 年的鱼群算法也属于群智能算法,鱼群算法没有高层指挥者,也不需要关于问题的背景和先验知识,多个人工鱼个体并行进行搜索,算法整体表现出快速向极值区域收敛的特性,具有较高的寻优速率。与传统的计算方法相比,群智能优化算法在搜索策略上比较突出的优点是,无集中控制、多代理机制、算法结构简单、隐含并行性、易理解和易实现。此外,将上述各种优化算法之间进行相互组合,在性能上取长补短,可以形成各种混合优化算法,也是算法研究及其应用研究的热点。对于本书研究的制造过程质量的智能诊断方法问题,进化算法和群智能算法适合用来解决神经网络和其他学习机的结构或者参数优化选择问题。

3.5.2 基于遗传算法和 LS-SVM 的质量诊断仿真实验

3.5.2.1 遗传算法及其实验设置

LS-SVM 用于解决模式识别问题时,其参数 γ 和 σ 理论上不存在确定的解析方法与支持向量机的性能相联系,因此本书使用遗传算法(Genetic Algorithm,GA)同时对这两个参数进行优化设计。遗传进化算法是模拟生物界自然选择和遗传机制的随机搜索算法,该算法使用概率变换规则而不使用确定性的变换规则在解空间寻优,反复将选择算子、交叉算子和变异算子作用于种群,在大多数情况下可以得到问题的最优解和近似最优解,且遗传算法不需要依赖于具体的待解决问题,搜索方向仅由适应度函数决定,因此适用于本书涉及的 LS-SVM 的参数优化问题。对于使用 LS-SVM 进行质量控制图模式识别问题,使用遗传算法需要考虑并确定以下几个问题。

1) 种群初始化

LS-SVM 的参数 γ 和 σ 的种群个体的数值范围原则上只要保证大于零即可,根据使用 LS-SVM 对于控制图模式分类的经验,将初始种群个体取值区间设置为 [0,30];参数 γ 和 σ 的每代种群大小影响到算法的收敛速度,根据经验将种群大小设置为 20 个个体,形成 20×2 的数据结构;对每个个体使用 10 位的二进制编码,个体经过编码成为染色体,参与寻优搜索计算。

2)适应度函数

适应度函数亦称为评价函数,其取值总是非负的,是用来度量群体中的个体在进化过程中有助于找到最优解的优良程度。适应度较高的个体被遗传到下一代的概率就大,反之,适应度小的个体被遗传到下一代的概率就要小些,因此,任何情况下总是希望适应度越大越好。个体的适应度就是遗传算法中自然选择的依据,是评价一个种群中的个体好坏的指标。

适应度函数的设计是为了指导遗传算法的搜索方向。适应度函数的设计依据主要是待解决问题所关心的指标。本书希望对六类控制图模式得到最大化的平均识别率,因此使用六类控制图的平均识别率(Average Recognition Rate,ARR)作为算法的适应度函数。

3)遗传操作之选择、交叉、变异

选择就是根据计算得到的个体的适应度给每个个体分配一个选择概率,其本质就是对种群中的个体进行优胜劣汰的操作,适应度较高的种群中的个体被遗传到下一代的概率较大。正确的选择操作既要避免有用遗传信息的丢失,又要提高全局收敛速度。选择操作不当,要么会造成个体的相似度过大使得进化停止,要么会导致适应度过大的个体误导整个群体的发展方向,使得进化失去多样性,产生进化早熟问题。选择操作的方式很多,常见的选择操作包括轮盘赌选择、无放回随机选择、随机竞争选择等。

交叉又称为重组,是把两个父个体的部分结构加以替换,以产生下一代新个体的过程,是遗传算法的产生子代的主要的和关键的操作,决定了遗传算法的全局搜索能力和算法的收敛速度。交叉操作的设计需要确定交叉点的位置和部分基因交换的方式两个问题。常见的交叉操作包括单点交叉、多点交叉、均匀交叉、循环交叉等。

变异操作就是以一个较小的概率对个体编码串上的个别位

置进行改变。交叉运算是产生新个体的辅助方法。合理的交叉运算可以改善遗传算法的局部搜索能力,防止遗传早熟现象。变异概率不宜取得过大,以免破坏优秀个体。变异包括均匀变异、基本位变异等多种形式。

本书的遗传算法参数优化试验中,以随机遍历抽样的方式进行选择,以 0.7 的交叉概率进行单点交叉,变异概率取 0.01,作为遗传终止条件的最大遗传代数根据经验设置为 50。

3.5.2.2 SVM 识别器设计

设计 LS-SVM 识别器所用各种的方法,包括训练/测试样本的生成方法与数量,多分类 LS-SVM 识别器的设计方法等,都与上一节相同,其仿真程序被当成子程序来调用,该子程序被命名为"控制图模式识别子程序"。

3.5.2.3 仿真实验过程

使用 LS-SVM 及遗传算法进行控制图模式识别仿真试验,包括参数寻优和模式识别性能测试两大部分,第二个部分的实验基本同上一节相同。仿真实验过程用图 3-5 表示,图中包含一个主程序流程图和一个子程序流程图。

3.5.2.4 遗传算法参数优化结果

使用 MATLAB7.4 与英国 Sheffield 大学的遗传算法工具箱,通过仿真实验得到 LS-SVM 的优化参数 $\gamma = 5.0733, \sigma = 8.2991$。图 3-6 给出了使用遗传算法经过 50 代遗传进化后的最优解(适应度值)变化曲线。

从图 3-6 可以看出遗传算法优化过程在第 37 代之后最优解趋于稳定。

3.5.2.5 控制图模式识别结果

使用通过上面遗传算法得到的优化参数 $\gamma = 5.0733, \sigma =$

8.2991，对 LS-SVM 模式识别器进行性能测试。得到的 LS-SVM 对六类控制图模式识别的性能结果如表 3-2 所示。

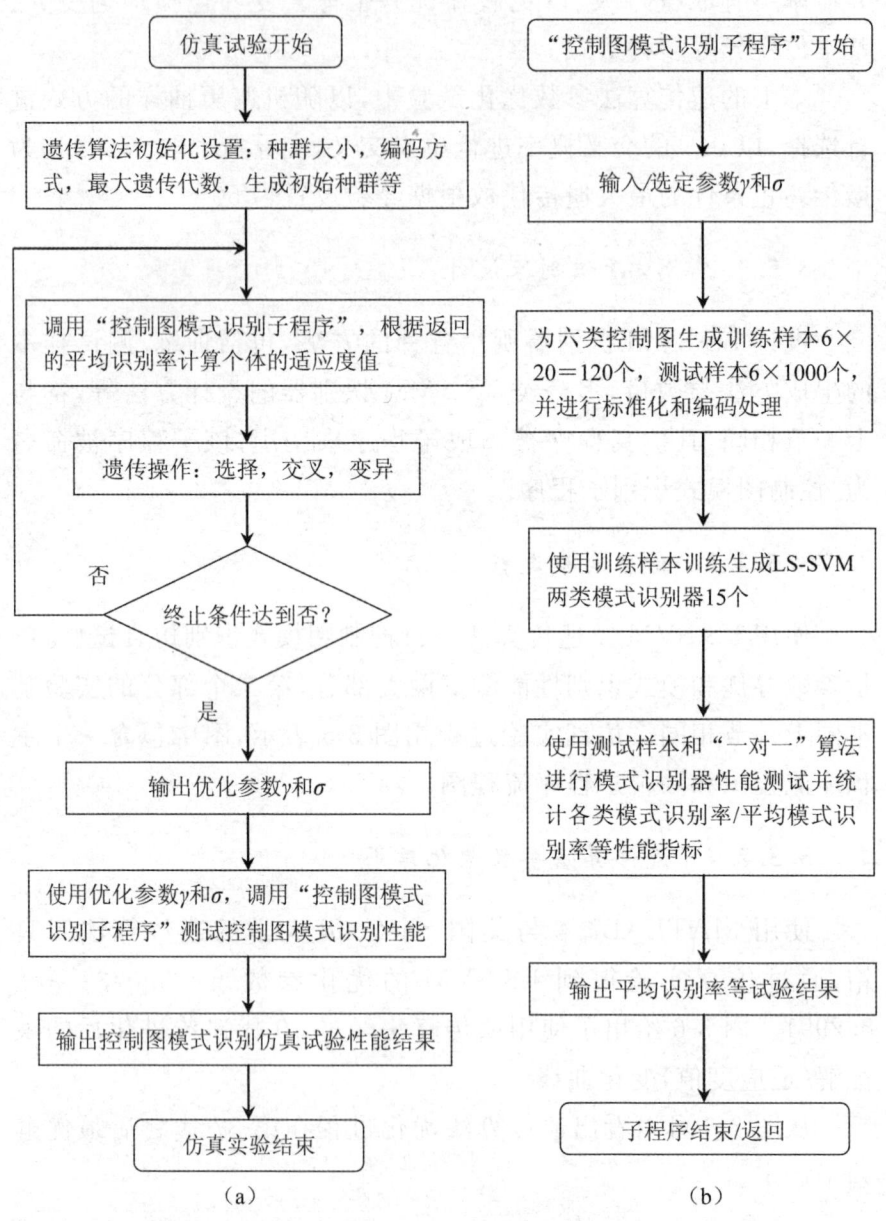

图 3-5 基于遗传算法和 LS-SVM 的控制图模式识别流程图
(a)仿真实验主程序；(b)仿真实验子程序

第3章 基于LS-SVM的小样本过程质量诊断

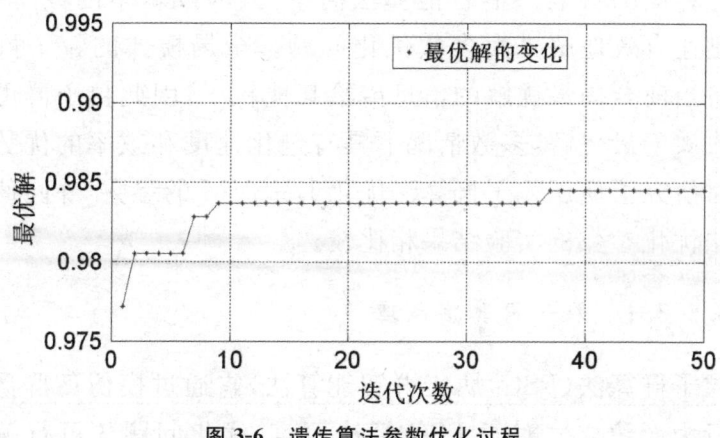

图 3-6 遗传算法参数优化过程

表 3-2 基于遗传算法和 LS-SVM 技术的控制图模式识别率(%)

模式类型	六类控制图模式识别率						平均识别率
	正常模式	向上趋势	向下趋势	向上阶跃	向下阶跃	周期模式	
本书结果	100	96.0	98.2	99.2	99.4	100	98.8
其他结果[53]	99.25	96.5	96.5	96.25	95.75	100	97.4

从表 3.5.1 的实验结果来看,通过使用遗传算法优化 LS-SVM 的参数,达到了较高的模式识别率。从整个实验过程来看,使用遗传算法优化 LS-SVM 的参数,不需要考虑过多的问题的背景,算法容易理解且设计简单,易于工程实现,对于 LS-SVM 两个参数的优化过程没有陷入局部最优解。但从解决 LS-SVM 控制图模式识别中的参数优化问题的仿真实验来看,遗传算法表现出的缺点就是参数优化过程收敛的速度慢,效率低。

3.5.3 基于粒子群算法和 LS-SVM 的质量诊断仿真实验

群智能算法是针对社会性群集动物的行为特征建模而得到的优化算法。社会性动物的个体行为看上去简单,但是,其群体的行为却表现出了高度智能的特征。蚁群算法和粒子群算法是群智能算法的代表。群智能算法简单,只涉及基本的数学运算,能够迅速有效地解决大多数优化问题。全局模式的粒子群算法可以将当前全局最优解的信息传给其他粒子,因此粒子群可以更快地收敛于最优解,多数情况下具有进化速度和效率的优势。本书同时研究了使用粒子群算法优化 LS-SVM 的参数,并同使用遗传算法优化参数的实验结果相比较。

3.5.3.1 粒子群算法原理

粒子群算法(PSO)属于群智能算法,它通过模仿鸟群觅食的社会行为而建立。对于 PSO 算法而言,优化问题的可行解对应于搜索空间中一只鸟的位置,称这些鸟为"粒子"。粒子群初始化时,随机产生优化问题的多个初始解,粒子群中的每个粒子都有一个由优化目标函数决定的适应度值,每个粒子还有一个决定下一代粒子飞行方向和距离的速度。搜索优化问题的最优解时,每个粒子同时跟踪自己找到的当前位置最优解(p_{iK})和整个种群找到的当前位置最优解(p_{gK}),并通过下面式子来更新自己的速度 v_K 和位置 x_K:

$$v_{K+1} = w v_K + c_1(p_{iK} - x_K) + c_2(p_{gK} - x_K) \quad (3.36)$$

$$x_{K+1} = x_K + v_{K+1} \quad (3.37)$$

对于 LS-SVM 识别器的参数优化问题,x_K 表示迭代过程中的待优化参数 $\{\gamma, \sigma\}$ 的当前取值。

3.5.3.2 仿真实验粒子群算法设计

1)粒子群算法的参数选取

式(3.36)中,c_1, c_2 为加速因子,取(0,2)区间均匀分布的随机数;每个粒子的最大限速取 $v_{max} = 2, v_K \in [-v_{max}, v_{max}]$;$w$ 为惯性因子,为了在全局搜索和局部搜索之间取得平衡以便减少进化迭代次数,w 的改进可以使用式(3.38),使得 w 在迭代过程中线性递减:

$$w = w^0 - (w^0 - w^\infty) \times K/(MS) \quad (3.38)$$

其中，w^0 为初始惯性因子，w^∞ 为终止惯性因子；M 为最大迭代次数，S 为递减系数。本书实验中取 $w^0=0.95, w^\infty=0.4, M=100, S=0.9$。

2）粒子的适应度设计

对于 LS-SVM 识别器需要优化的两个大于零的实数参数，正则化参数 γ 和核函数参数 σ，这两个参数的整体组合成为一个"粒子"参与进化迭代，由于研究重点是 LS-SVM 识别器的识别率性能，文中使用六类控制图的平均模式识别率性能作为粒子的适用度值。

3）种群初始化

种群的初始化涉及两个方面：位置和速度。本书试验中每代的种群容量设为 10；形成初始粒子种群时，γ 和 σ 采用实值编码方法，粒子的速度和位置的初始取值范围根据经验设定在区间（0，20）之中随机取值。

3.5.3.3 LS-SVM 识别器设计

设计 LS-SVM 识别器设计所用的技术，包括训练/测试样本的生成，LS-SVM 识别器的设计方法等，都与上一节相同。

3.5.3.4 仿真试验过程

使用 LS-SVM 及 PSO 算法进行控制图模式识别仿真试验，包括参数寻优和模式识别性能测试两大部分，仿真过程的步骤如下。

(1) 随机产生 10 个 $\{\gamma, \sigma\}$ 作为初始粒子群的位置，并用相同的方法初始化粒子速度；调用"控制图模式识别子程序"计算当前初始粒子群的每个粒子的适应值（模式识别率），作为初始粒子群的最优个体适应值，并计算初始粒子群的全局最优位置和初始全局最优适应值。设置记忆当前粒子群每个个体最优位置的变量、记忆全局最优位置的变量，设置记忆当前粒子群每个个体最优适应值的变量、记忆全局最优适应值的变量。

(2)使用式(3.36)、式(3.37)、式(3.38)更新每个粒子的速度和位置,形成新一代粒子群 $\{\gamma,\sigma\}$;对于新一代粒子群,调用"控制图模式识别子程序"计算新一代种群中每个粒子的适应值。

(3)根据新一代粒子群适应度值的计算结果,更新每个粒子个体所对应的最佳位置 p_{iK} 和相应的最佳适应值;更新粒子种群所对应的全局最佳位置 p_{gK} 和全局最佳适应值。

(4)重复步骤(2),直到满足最大迭代次数,输出得到的优化参数 $\{\gamma,\sigma\}_{best}$。

(5)使用优化参数 $\{\gamma,\sigma\}_{best}$,调用"控制图模式识别子程序"进行 LS-SVM 方法控制图模式识别的性能测试,并记录试验结果。

上面步骤中的"控制图模式识别子程序"的程序框图参见图 3-5。

3.5.3.5 粒子群算法参数优化结果

本书使用 MATLAB7.4 软件工具,使用粒子群算法和 LS-SVM 识别器进行控制图模式识别仿真实验,通过 PSO 算法得到 LS-SVM 识别器的优化参数 $\gamma=19.6154, \sigma=8.4538$。图 3-7 表示是经过 50 次迭代时全局最优适应度值的变化曲线。

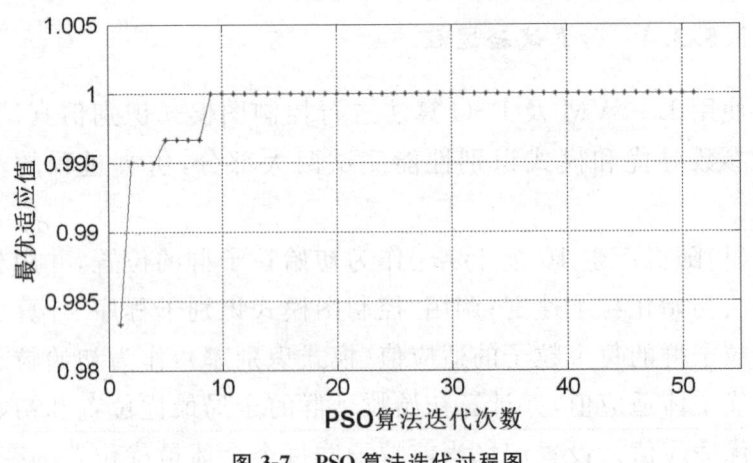

图 3-7 PSO 算法迭代过程图

图 3-7 中,第一个点表示在初始时刻 $t=1$ 时的全局最优适应

值（最优解），整个进化过程仅仅通过 8 次迭代，在 $t=9$ 时刻最优适应值已经趋于稳定，表现出了较快的进化速度。

3.5.3.6 LS-SVM 模式识别结果

使用得到的优化参数 $\gamma=19.6154, \sigma=8.4538$，测试得到的 LS-SVM 的模式识别性能结果见表 3-3。

表 3-3 LS-SVM 控制图模式识别率（%）

模式类型	六类控制图模式识别率						平均识别率
	正常模式	向上趋势	向下趋势	向上阶跃	向下阶跃	周期模式	
本书 PSO+LS-SVM 的结果	97.8	99.4	99.4	99.6	99.8	99.8	99.3
仅使用 LS-SVM 的结果	97.2	99	99.5	96.2	93.3	97.4	97.1
PCA+SVM 方法[53]的结果	99.25	96.5	96.5	96.25	95.75	100	97.4

由试验结果可以看出，经过粒子群算法参数优化的 LS-SVM 识别器对于控制图模式的识别结果，相比没有使用粒子群算法的其他的控制图模式识别的结果，LS-SVM 识别器对于控制图模式的识别性能都有提高。虽然目前对于 PSO 算法的内部机理研究尚未成熟，对其参数的设置和构造没有实质性的认识，但从使用 PSO 算法优化 LS-SVM 识别器参数的实验过程和结果来看，使用粒子群算法优化 LS-SVM 参数的速度相比使用遗传算法优化的速度大大加快，表现出了更高的使用效率，应当成为优化神经网络和学习机结构与参数时优先考虑使用的方法。

3.6 本章小结

先进制造环境下的生产过程具有多品种小批量的特征，在小样本条件下对制造过程进行质量诊断是市场环境的客观要求；同时，质量管理必须满足经济性要求。因此研究过程质量诊断技术必须使用尽可能少的测试样本,为制造企业节约质量成本。

本章首先介绍了小批量条件下解决质量诊断问题的 SPC 控制图方法,继而提出使用基于统计学习理论的支持向量机技术来实现在有限样本条件下的控制图模式识别,并对统计学习理论和支持向量机技术做了介绍。为验证所提出方法的效果,设计了 LS-SVM 模式识别器,并通过仿真实验研究和评估其性能。实验结果表明,LS-SVM 模式识别技术是小样本质量诊断的有效方法。另外,针对 LS-SVM 技术本身存在的参数优化问题,提出使用进化算法来优化选择 LS-SVM 识别器的参数,并分别使用具有代表性的遗传算法和粒子群算法,通过仿真实验来优化 LS-SVM 识别器的参数,并将实验结果做了比较研究,证明了提出方法的有效性。

第 4 章 基于 Cuscore 统计量的过程质量智能诊断

传统的控制图理论(Shewhart 图、CUSUM 图、EWMA 图等)一般假定过程中的异常变化是"不可预期的"(unexpected),过程属于白噪声过程[139,140]。然而,对于许多制造过程而言,导致过程异常的原因和过程异常的很多现象在很多情况下却是"可预期的"(anticipated)。制造过程经过一段时间的运行,现场工程师或操作员工通常会对过程的异常现象进行总结,从而形成关于过程异常的先验知识,从而使得过程异常具有一定的"可预期性"。这种可预期性往往体现了制造过程本身的规律性。在进行过程诊断的研究中,如果充分利用积累下来的关于过程的先验知识,无疑会提高过程诊断的效率。

本书针对制造过程中的预期异常信号的诊断问题,研究了 Cuscore 图对于非线性二次预期异常信号的检测性能;针对标准 Cuscore 图中存在的"失配"问题,提出了使用移动窗口法和 LS-SVM 模式识别技术进行变点检测的模型,解决了标准 Cuscore 图中存在的"失配"问题,提高了 Cuscore 图技术对于预期异常信号的诊断效率。

4.1 Cuscore 统计量与过程异常诊断问题

伴随着制造企业的发展,企业会根据以往对过程的操作经验不断积累关于过程故障与质量异常的经验知识,包括异常信号的类型、异常信号的幅值及其经验函数表达式等,并且积累与该故障相对应的故障的部位、原因和故障处理方法等经验知识,以便当过程中再次出现这类故障信号特征已知的"预期异常信号"时,可以对其进行有效诊断。但是,传统控制图在监控过程时并没有充分利用有关过程异常的经验知识,使得传统控制图在检测诊断"预期异常信号"方面存在着缺陷。由于"预期异常信号"反映了制造过程的本质特征[140],因此寻找新的质量诊断工具以便充分利用有关过程异常的先验知识,提高过程诊断的效率,是质量诊断研究中的一个重要内容。

统计学中的Cuscore(Cumulative Score)统计量由Fisher[141]在1925年提出,Box[142]在Cuscore统计量的基础上提出了Cuscore控制图(Cuscore Chart)。Cuscore控制图[157-160]尤其适用于诊断由于预期异常信号而造成的过程异常,同时,还可以有效利用企业以往积累下来的工作经验,从而提高过程诊断的效率[143]。因此,有必要重新研究Cuscore统计量在过程质量诊断与改进中的作用,使之有效地与其他管理模式相整合,充分发挥其在过程诊断方面的效力。

4.1.1 Cuscore统计量

Box和Nembhard[142,143]等人基于统计学中似然比检验(likelihood ratio test)概念和如下的统计过程模型来建立Cuscore统计量,

$$a_i = a_i(Y_i, X_i, \gamma), i = 1, 2, \cdots, t \tag{4.1}$$

在公式(4.1)中,Y_i为所关心的过程质量特性的输出观测值,X_i为过程的输入可控变量,γ是关于待诊断的预期异常信号$f(t)$的某个未知参数(如,阶跃信号的幅度,斜坡信号的幅度等;γ_0为γ的初始值),a_i为白噪声信号,即$a_i \sim N(0, \sigma_a^2)$,方差$\sigma_a^2$为已知并且不依赖于参数$\gamma$,不失一般性,研究中可假定$\sigma_a^2 = 1$。在模型(4.1)中,当过程中不存在预期异常信号时,过程中仅存在白噪声a_i。当过程运行一段时间之后,采用测量的方法可以得到一系列

过程的观测数据,此时,对于 γ 的每一个取值,使用式(4.1)可以计算得到 a_i 的一系列取值,那么,噪声样本 a_i 的对数似然函数可以使用下式进行计算:

$$L = -\frac{1}{2\sigma^2}\sum_{i=1}^{t}a_i^2 + c \tag{4.2}$$

式(4.2)中,c 是不依赖于 γ 的常数。对上式在 $\gamma=\gamma_0$ 处进行微分运算,并且令 $d_i = -\left.\frac{\partial a_i}{\partial \gamma}\right|_{\gamma=\gamma_0}$,可得:

$$\left.\frac{\partial L}{\partial \gamma}\right|_{\gamma=\gamma_0} = \frac{1}{\sigma^2}\sum_{i=1}^{t}a_{i0}d_i \tag{4.3}$$

Fisher 将累计得分(Cuscore)统计量定义为:

$$Q_t = \sum_{i=1}^{t}a_{i0}d_i \tag{4.4}$$

式中,a_{i0} 为 a_i 的初始值(a_i 在 $\gamma=\gamma_0$ 时的值)。d_i 通常被称为"探测算子"(detector),表示残差 a_i 对于被监测参数 γ 的微分值,即残差 a_i 对于参数 γ 的变化率。可见,Cuscore 统计量就是过程残差 a_{i0} 与残差 a_{i0} 相对于过程参数 γ 的变化率(d_i)的乘积的累计和函数。当绘制 Cuscore 图时,Q_t 表示 Cuscore 图上在时间 t 的值。

当使用 Cuscore 图监控过程质量特性的正/负向变化时,可以使用下式中的 Q_t^+/Q_t^- 统计量:

$$\begin{cases} Q_t^+ = \max(0, Q_{t-1}^+ + a_{i0}d_t) \\ Q_t^- = \min(0, Q_{t-1}^- + a_{i0}d_t) \end{cases} \tag{4.5}$$

其中,$Q_0^+ = Q_0^- = 0$。该 Cuscore 图的工作原理是,当 Q_t^+/Q_t^- 统计量超出上/下控制限,Cuscore 图发出报警信号,由此判断过程失控。如果 Cuscore 统计量在控制限之内,认为过程受控,则需要通过继续计算 Cuscore 统计量来监控过程的状态。

4.1.2 过程模型

对于一个制造过程,如果其质量特性观测数据是独立的,可以使用经典控制图直接对该质量特性的原始数据进行监控。如果质量特性观测数据是自相关的,对质量观测数据直接使用经典控制图会导致过多的误报,此时 SPC 技术中经常使用时间序列模型(time series model)和残差控制图(residual-based chart)来实现过程异常监控[144]。通过使用时间序列模型对自相关过程数据充分建模,其残差(即,实际观察值与回归估计值的差)常常是统计近似独立的,因此可以对残差数据使用经典控制图来达到监控过程异常的效果。

由于制造过程本身的特点(如流程性过程),以及采样时间间隔等因素的影响,制造过程的质量特性观测值经常具有自相关的特征。自相关过程数据常常使用 ARMA(Autoregressive Moving Average)来描述。对于使用 ARMA 模型来描述的自相关过程数据,Cuscore 图是另外一种对于自相关过程中出现的预期异常信号进行监控的重要方法。本书研究中使用的开环时间序列模型如图 4-1 所示,该模型可以很好地对应于制造过程中某些单一制造工序的情形。

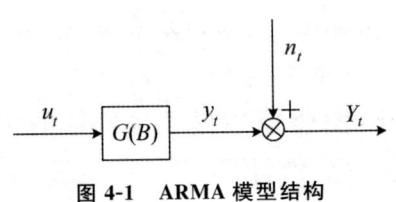

图 4-1　ARMA 模型结构

图 4-1 中的模型包括两个部分:动态过程和扰动过程。$G(B)$ 是动态过程的传递函数,其输入为 u_t,输出为 y_t。当假定模型中不存在扰动 n_t 时,动态过程的稳态输出 Y_t 的目标值设定为 T,不失一般性,研究中可以令 $T=0$。扰动 n_t 由预期异常信号和使用 $ARMA(p,q)$ 模型表示的有色噪声 $\frac{\theta(B)}{\phi(B)}a_t$ 组成,a_t 如前所述,为白噪声。即,$n_t = \frac{\theta(B)}{\phi(B)}a_t + \gamma f(t)$,$B$ 称为"后向移位算子",满足 $B^K X_t = X_{t-K}$,$\phi(B)$ 和 $\theta(B)$ 分别表示"自回归,AR"和"移动平

第 4 章 基于 Cuscore 统计量的过程质量智能诊断

均,MA"多项式,即 $\phi(B) = 1 - \phi_1 B - \phi_2 B^2 - \cdots - \phi_p B^p$,$\theta(B) = 1 - \theta_1 B - \theta_2 B^2 - \cdots - \theta_q B^q$。$f(t)$ 表示预期异常信号的性质,γ 表示预期异常信号的某个参数。显然,当 $\phi(B) = 1$,$\theta(B) = 1$ 时,图 4-1 中的模型表示的是白噪声过程。

白噪声过程可以看成为 ARMA 过程的一个特例。Box 和 Nembhard 的研究结果指出[145,146],当 Cuscore 图用于检测白噪声过程中的尖峰型预期信号(Spike Signal)时,其效果等同于传统休哈特控制图;当 Cuscore 图用于检测白噪声过程中的阶跃型预期信号(Step Signal)时,其效果等同于 CUSUM 图;当 Cuscore 图用于检测白噪声过程中的指数型预期信号(Exponential Signal)时,其效果等同于传统 EWMA 控制图;当 Cuscore 图用于检测白噪声过程中的凸块型预期信号(Bump Signal)时,其效果等同于传统算数移动平均(AMA,Arithmetic Moving-Average)控制图。

当动态过程处于稳态并且过程中没有出现预期异常信号时,图 4-1 中的模型被称为"零模型"(Null Model),表示为:

$$Y_t = T + \frac{\theta(B)}{\phi(B)} a_{t0} \qquad (4.6)$$

式(4.6)中假定 a_t 的值可以为 0,即 a_{t0} 的取值仅仅是残差。

当动态过程处于稳态并且过程中在某个已知时刻 τ(不妨令 $\tau = 1$)出现了预期异常信号时,则图 4-1 中的模型被称为"差异模型"(Discrepancy Model),可以表示为:

$$Y_t = T + \frac{\theta(B)}{\phi(B)} a_t + \gamma f(t) \qquad (4.7)$$

假定 $\theta(B)/\phi(B)$ 可逆,对式(4.6)两端对参数 γ 求导数,可得探测算子 d_i:

$$d_i = -\frac{\partial a_i}{\partial \gamma}\bigg|_{\gamma = \gamma_0} = f(t) \frac{\phi(B)}{\theta(B)} \qquad (4.8)$$

将式(4.8)、式(4.6)代入式(4.4)中可得图 4-1 中模型所示的 Cuscore 统计量为:

$$Q_t = \sum_{i=1}^{t} f(t) \frac{\phi(B)}{\theta(B)} a_{t0} \qquad (4.9)$$

需要指出,式(4.9)中假定了预期异常信号出现的时刻 τ 为已知并且 $\tau=1$,该情况类似于制造过程中出现的有计划成批更换物料等情形。但是对于一个实际的制造过程,如果制造过程中预期异常信号出现的时刻 τ 未知,则探测算子 d_i 和残差 a_i 之间存在一个所谓"失配"(mismatch)问题。d_i 和 a_i 之间的失配(即,不同步),会导致 Cuscore 统计量对于预期异常信号的检测性能下降。

4.2 Cuscore 统计量对于非线性预期异常信号的诊断性能

对于白噪声过程,使用传统的休哈特图、EWMA 图、CUSUM 图对过程测量数据直接进行监控,可以取得良好的过程异常诊断效果。对于自相关过程数据,使用传统控制图直接对过程测量数据进行监控会导致过多的误报,此时可以考虑使用所谓"基于残差的控制图"方法,即,利用时间序列模型对过程原始数据进行建模,然后对残差数据使用传统控制图进行异常监控[147]。对于存在自相关现象的过程,Cuscore 控制图尤其适用于诊断故障特征(Fault Signature)已知的预期异常信号,因此有必要研究 Cuscore 图在该情况下对过程异常进行诊断的性能,以便发挥其在过程监控方面的作用。

4.2.1 非线性二次预期信号及其 Cuscore 图的绘制

实际的工业制造过程中出现的预期异常信号是复杂和多种多样的,除了典型的阶跃(Step)、尖峰(Spike)、凸块(Bump)等信号外,还有许多其他类型的预期异常信号,这些异常信号更多地具有非线性特征。Nembhard 等人[143]使用 ARMA(1,1)过程模型和 Cuscore 图对阶跃(Step)、凸块(Bump)等典型预期异常信号的检测能力进行了研究评价,本书针对更具一般性的非线性二次预期信号进行仿真研究,评价 Cuscore 图对该类异常信号的诊断性能。本书提出的非线性二次预期信号的类型和性质使用了下面的表达式:

$$f(t) = \begin{cases} (t-\tau)^2, t > \tau \\ 0, t \leqslant \tau \end{cases} \quad (4.10)$$

式(4.10)中的表达式 $f(t)$ 只是表示了预期信号的类型和性质。设 γ 表示可以有不同取值的参数,则表达式 $\gamma f(t)$ 中包含了该非线性二次预期信号的幅度变化快慢方面的信息。考虑到工业过程的单个工序多数属于一阶或二阶动态噪声过程[148],本书研究针对图 4-1 中描述的 ARMA(1,1)模型,选取 $p = q = 1$;同时为简单起见,在绘制和评价 Cuscore 图对预期异常信号的检查性能时,假定预期异常信号的出现时刻 τ 是已知的,设定为 $\tau=1$。

绘制 Cuscore 图需要确定该控制图的控制限。Cuscore 图的控制限可以根据"控制图的受控 ARL 为 370"的概念使用下面的步骤通过仿真的方法来确定:

(1)生成 1000 个点的白噪声时间序列信号(即 1000 点的 a_{t0} 观测值)。

(2)使用公式(4.9)计算 Cuscore 统计量,可以得到一张 1000 点的 Cuscore 图。

(3)重复步骤(1)、(2)一万次,得到一万张 Cuscore 图。选择使得这一万张控制图的受控平均链长 ARL 为 370 的控制限作为 Cuscore 图的控制限。

使用上述 Cuscore 图控制限绘制方法,选择图 4-1 的 ARMA (1,1)模型的参数 $\phi_1 = 0.8, \theta_1 = -0.2$ 及非线性二次预期异常信号参数 $\gamma=1$,使用 MATLAB 软件进行蒙特卡罗(Monte-Carlo)仿真得到对线性二次预期异常信号进行异常监控的 Cuscore 图如图 4-2 所示。为了比较 Cuscore 图与标准休哈特控制图对于预期异常信号的检测能力,本例中同时绘制出使用标准休哈特控制图对于提出的非线性二次预期异常信号的检测情况,每张 Cuscore 图和标准休哈特图都包含 30 个数据点。

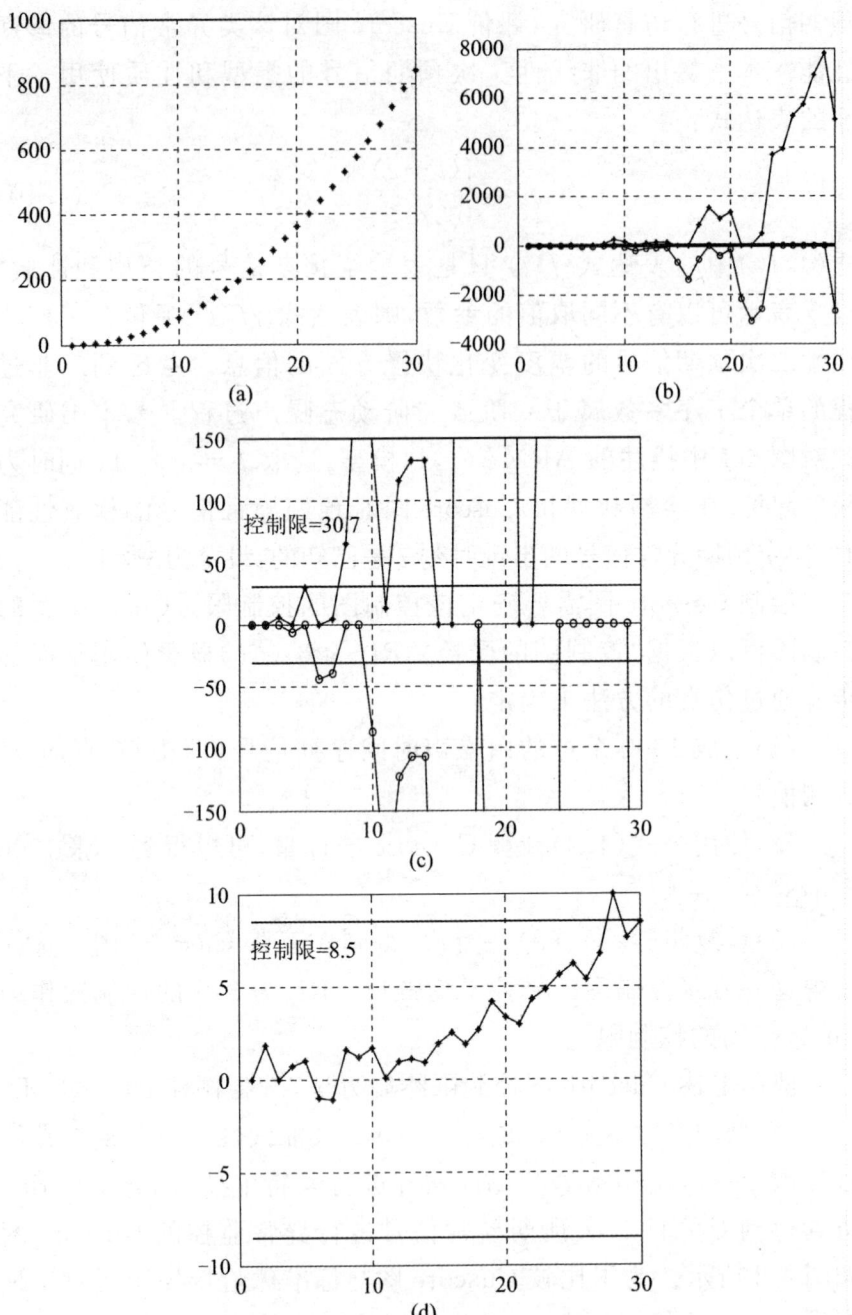

图 4-2 非线性二次预期异常信号及其 Cuscore 图、休哈特图

(a)预期异常信号;(b)Q^+,Q^-统计图;

(c)带控制限的(Q^+,Q^-)双侧 Cuscore 图;(d)休哈特控制图

第 4 章 基于 Cuscore 统计量的过程质量智能诊断

由图 4-2(c)看出,通过仿真方法得到的 Cuscore 图的控制限为 30.7,Cuscore 图在 $t=6$ 时就对非线性二次预期异常信号发出了异常报警信号,但是如果使用标准休哈特图对过程观测数据进行异常诊断,根据 3sigma 控制限准则通过计算得到的控制限为 8.5,图 4-2(d)直到 $t=28$ 时才发出异常报警信号。

4.2.2 识别率及识别速度

评价 Cuscore 图的性能,可以使用的指标包括准确率(Accuracy Rate)和失控平均链长(Average Run Length,ARL),它们分别被用来评价 Cuscore 图对于预期异常信号的总体检测能力和检测速度。本书研究中对于图 4-1 表示的 ARMA(1,1)模型参数 ϕ_1 和 θ_1,以及预期异常信号的参数 λ 选取不同的取值,研究 Cuscore 图对于式(4.10)表示的二次非线性预期异常信号的检测准确率和识别速度(失控 ARL),并与使用休哈特图对该二次非线性预期异常信号进行异常检测时的性能表现相比较。通过 1000 次仿真试验得到的实验结果见表 4-1 和表 4-2。这两个表中的括号中的数据为休哈特图的性能数据。

表 4-1 Cuscore 图及休哈特控制图的检测正确率(%)(括号内为休哈特图的数据)

λ	$\phi_1=-0.8$ $\theta_1=0.2$	$\phi_1=-0.6$ $\theta_1=0.4$	$\phi_1=-0.4$ $\theta_1=0.6$	$\phi_1=-0.2$ $\theta_1=0.8$	$\phi_1=0.8$ $\theta_1=-0.2$	$\phi_1=0.6$ $\theta_1=-0.4$	$\phi_1=0.4$ $\theta_1=-0.6$	$\phi_1=0.2$ $\theta_1=-0.8$
0.01	100 (53.2)	100 (60.8)	100 (68.1)	100 (67.6)	100 (47.3)	100 (51.8)	100 (53.4)	100 (55.9)
0.02	100 (82.4)	100 (86.3)	100 (90.7)	100 (88.0)	100 (62.3)	100 (64.2)	100 (70.1)	100 (68.8)
0.03	100 (93.2)	100 (95.9)	100 (96.8)	100 (97.3)	100 (71.4)	100 (78.2)	100 (80.6)	100 (84.0)
0.04	100 (98.4)	100 (99.2)	100 (97.8)	100 (99.1)	100 (82.0)	100 (86.4)	100 (86.10)	100 (89.3)

表 4-2 Cuscore 图和休哈特控制图的失控 ARL(括号内为休哈特图的数据)

λ	$\phi_1=-0.8$ $\theta_1=0.2$	$\phi_1=-0.6$ $\theta_1=0.4$	$\phi_1=-0.4$ $\theta_1=0.6$	$\phi_1=-0.2$ $\theta_1=0.8$	$\phi_1=0.8$ $\theta_1=-0.2$	$\phi_1=0.6$ $\theta_1=-0.4$	$\phi_1=0.4$ $\theta_1=-0.6$	$\phi_1=0.2$ $\theta_1=-0.8$
0.01	4.04 (27.7)	3.77 (28.8)	4.35 (27.8)	5.4 (27.8)	2.9 (26.5)	2.8 (27.4)	3.0 (27.7)	3.2 (27.8)
0.02	4.25 (28.2)	4.11 (28.3)	4.31 (28.4)	5.6 (28.4)	2.9 (28.1)	2.9 (28.2)	3.2 (28.3)	3.5 (28.3)
0.03	4.47 (28.5)	4.22 (28.5)	4.41 (28.5)	5.0 (28.5)	3.0 (28.4)	3.0 (28.5)	3.1 (28.5)	3.2 (28.5)
0.04	3.96 (28.5)	4.38 (28.6)	4.46 (28.7)	4.5 (28.6)	2.8 (28.5)	3.1 (28.6)	3.1 (28.6)	3.5 (28.7)

根据表 4-1 和表 4-2 试验结果可知,对于本书研究中提出的二次非线性预期异常信号 $f(t)$,Cuscore 图可以实现 100% 检测,并且 Cuscore 图对于该异常信号的检验能力不受参数 λ 取值变化的影响;而传统休哈特图对于该异常信号的检测能力受到参数 λ 的取值以及有色噪声模型参数 ϕ_1 和 θ_1 的取值的影响较大,休哈特图对于预期信号幅度变化较小(即 λ 较小)的异常情况的检测正确率明显低于 Cuscore 图。在对预期异常信号的检测速度性能方面,根据失控 ARL 可知,Cuscore 图对于提出的预期异常信号 $f(t)$ 的检测速度明显高于传统休哈特图。

4.3 解决 Cuscore 图失配问题的智能变点模型

4.3.1 Cuscore 图的失配问题

在过程数据呈现有色噪声特征的制造环境中,Cuscore 图对于预期异常信号的诊断比传统控制图更加迅速有效。将 Cuscore

图应用于过程质量监控,可以有效利用在以往生产中积累的关于过程异常的知识和处理经验,从而提高过程质量诊断与异常处理的效率。但是,由本书 4.1 节可知,目前的 Cuscore 技术中需要假定预期异常信号的出现时刻为已知的,这种假设在实际的制造过程中是存在并且合理的,比如,制造过程更换物料常常会导致过程的某个质量特性产生类似阶跃信号的异常变化,由于更换物料的时间是已知的,因此可以假定该阶跃预期异常信号的出现时刻是已知的。但是,对于实际的制造过程,很多情况下预期异常信号的出现时刻是未知的,这时使用 Cuscore 图技术监控过程无法保证预期异常信号与 Cuscore 统计量的检测算子之间的同步,即存在所谓失配问题。Cuscore 图失配问题的实质就是无法保证 Cuscore 图的触发时刻与异常信号的出现时刻之间的同步,该问题会导致 Cuscore 图对于预期异常信号的监控能力下降。针对此问题,本节的内容研究如何使用 LS-SVM 模式识别技术、利用统计学中变点检测的概念来确定预期异常信号的出现时刻,从而确定 Cuscore 图的触发时刻,提高 Cuscore 图对于预期异常信号的检查性能。

4.3.2 触发 Cuscore 图的智能模型

在应用 Cuscore 图的过程中,如果根据先验经验,预期异常信号的出现时刻 τ 是已知的(不失一般性,可以假定 $\tau=1$),那么可以在时刻 τ 触发 Cuscore 图以便诊断过程异常,这种情况下的 Cuscore 图可以称之为"标准 Cuscore 图"。然而,实际的制造过程存在更为复杂的情况,因为制造工程师常常无法预测和确定预期异常信号的出现时刻,此时也就无法保证 Cuscore 图的初始化时间同预期异常信号的出现时间之间保持一致,该问题即是 Cuscore 图技术中的失配问题。失配问题会导致 Cuscore 图的性能下降。显然,当预期异常信号的真实出现时刻 τ 为未知时,为了提高和改善 Cuscore 图对于预期异常信号的检测性能,如果可以对预期异常信号出现的真实时刻 τ 进行辨识以便得到 τ 的估计值 $\hat{\tau}$，就可以在时刻 $\hat{\tau}$ 触发（即，初始化）Cuscore 图，使得 Cuscore 图的触发时刻最大限度接近于预期异常信号的的真实出现时刻 τ，从而提高 Cuscore 图对于预期异常信号的检测性能。为此，本书在图 4-1 的基础上，提出使用 LS-SVM 模式识别技术来寻找 Cuscore 图的触发时刻 $\hat{\tau}$ 的智能方法，其模型结构和工作原理如图 4-3 所示。

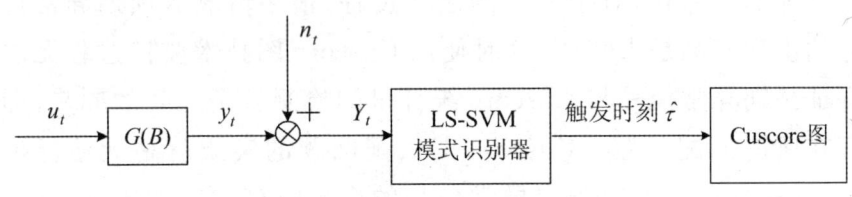

图 4-3　Cuscore 图的智能触发模型

在图 4-3 的模型中，制造过程的开始运行时刻设定为 $t=0$。随着过程的运行，可以得到过程质量特性的观测数据，这些数据首先要使用 LS-SVM 模式识别器进行分析。当 LS-SVM 模式识别器在某个时刻（将该时刻表示为 $\hat{\tau}$）首次发出异常模式报警信号，就将该时刻 $\hat{\tau}$ 看作是"最有可能"的预期异常信号的出现时刻，因此就在该时刻触发（初始化）Cuscore 图。这里期望经过 LS-SVM 模式识别器辨识得到的预期异常信号的出现时刻 $\hat{\tau}$ 尽可能接近预期异常信号实际出现的时刻 τ。换言之，从 LS-SVM 模式识别器首次发出异常报警的时刻 $\hat{\tau}$ 起，LS-SVM 模式识别器不再继续起作用，而 Cuscore 统计量 Q_t 在时刻 $\hat{\tau}$ 被触发并开始进行过程质量异常监控，当 Cuscore 统计量 Q_t 超过控制限，Cuscore 图发出过程异常报警信号。应当看到，在图 4-3 的模型中，LS-SVM 模式识别器并不是用来对预期异常信号发出报警，其作用仅仅是为 Cuscore 图提供一个触发时间 $\hat{\tau}$。

为了验证图 4-3 提出的模型的有效性，我们将对其工作性能进行评估。考虑到均值漂移（Mean Shift）信号是制造过程中最为常见的预期异常信号，因此在图 4-3 中选用 ARMA(1,1) 模型和均值阶跃类型的信号作为预期异常信号来进行相关的研究。

4.3.3 变点检测的移动窗口法

从上面的讨论可以看出,上节中涉及的确定预期异常信号出现时刻的问题,实际上就是统计学中的变点问题[149-151]。对于制造过程的质量特性的观测值,按其出现时间的先后和空间位置进行排列,就会形成质量特性的时间序列数据。当过程中存在的某个已知或者未知因素使得时间序列样本数据在某个时刻或者位置发生突然变化,这个时刻或者位置就是所谓"变点"。变点可以认为是模型中的某个或某些量起突然变化之点。从统计学的角度,变点的出现主要表现在过程的分布或者其数字特征(均值、方差等)发生变化。相应地,变点分析包括分布变点、均值变点和方差变点等。变点问题在统计学中有多方面的研究内容,其中的一个重要的研究内容就是在过程的时间序列数据中确定数据发生了异常变化的位置(或者发生异常变化的时刻)。制造过程质量特性数据中出现变点,意味着过程的质量特性发生了某种质的变化。

4.3.3.1 移动窗口法

统计学分析和研究时间序列数据中存在的变点问题,其实质就是对时间序列进行分类。即,首先对总的时间序列数据进行分段,形成多个时间序列子序列,然后通过假设检验的方法来判断每段时间序列子序列性质的变化,从而确定变点的存在。迄今为止,对于变点问题的研究基本上都是使用统计学的方法[152-156]。但是,使用统计学的方法研究变点问题需要假定或者知道关于过程的分布以及参数等先验知识,但是这些假设在实际的制造过程中往往难以真正满足或者验证,这使得研究变点问题的统计学方法在实际应用中存在着一定的困难。我们在研究中提出使用"移动窗口法"和基于 LS-SVM 的模式识别技术来确定时间序列数据中变点的出现位置/时刻,并以此为根据来触发 Cuscore 图,从而减少失配问题对于使用 Cuscore 图检测预期异常信号的不良影响。

对于一个连续运行的制造过程,可以获得其时间序列观测数据,表示为 $Y_1, Y_2, \cdots, Y_n, \cdots$。为了分析过程数据中的变点,可以使用移动窗口和模式识别技术来分析处理过程时间序列观测数据。设移动窗口的长度(窗口大小)为 l_w,移动窗口每次移动的步长为 h_m,当移动窗口经过 d 次移动之后,由过程的时间序列数据(即 $Y_1, Y_2, \cdots, Y_n, \cdots$)可以构造出如下的时间子序列数据,

$$\begin{cases} Y'_1 = \{Y_1, Y_2, \cdots, Y_{l_w}\} \\ Y'_2 = \{Y_{1+h_m}, Y_{2+h_m}, \cdots, Y_{l_w+h_m}\} \\ \vdots \\ Y'_d = \{Y_{1+(d-1)h_m}, Y_{2+(d-1)h_m}, \cdots, Y_{l_w+(d-1)h_m}\} \\ \vdots \end{cases}$$

$$d = 1, 2, 3, \cdots \tag{4.11}$$

为了使用 LS-SVM 模式识别技术来判定变点的出现,一个时间子序列 Y'_d 数据被定义为一个模式数据来处理。时间子序列 Y'_d 数据的正常模式和异常模式需要预先加以研究和定义。对于时间子序列 Y'_d 的正常模式,其数据应当仅仅包含受控过程的数据,而不能包含预期异常信号数据;对于时间子序列 Y'_d 的异常模式,其数据不仅包含受控过程的数据,还应当包含预期异常信号数据。通过对时间子序列 Y'_d 的模式进行分类和分析,就可以找到预期异常信号的出现时刻。当时间序列 Y'_d 由正常模式变为异常模式,就可以判定预期异常信号出现在这个异常模式的移动窗口中。

相比于统计学方法,本书提出的使用移动窗口法和 LS-SVM 模式识别技术来分析变点问题的方法具有其自身的优势,即基于 LS-SVM 模式识别的人工智能技术不依赖于关于过程的分布等

先验知识,因此具有更大的应用范围。

4.3.3.2 LS-SVM 模式识别器的训练

在使用移动窗口和模式识别技术来确定预期异常信号的出现时刻之前,需要对 LS-SVM 模式识别器进行训练,使其具有要求的模式识别能力。训练和测试 LS-SVM 模式识别器的样本仍然通过 Monte-Carlo 仿真的方法来获取。根据图 4-1 表示的 ARMA(1,1) 动态过程参数模型,在仿真实验研究中,正常模式的时间子序列 Y'_d 训练/测试数据集可以使用公式(4.6)仿真得到,异常模式的时间子序列 Y'_d 训练/测试数据可以使用公式(4.7)仿真得到。生成训练数据的仿真实验需要在不同的实验参数条件下进行,包括预期异常信号参数 γ、移动窗口长度 l_w 以及 ARMA(1,1)模型参数 ϕ_1 和 θ_1。对于均值阶跃类型的预期异常信号,在设计使用移动窗口来表示的异常时间序列模式数据时,不失一般性,我们在研究中将阶跃点设计在时间序列数据的中点。均值阶跃类型的预期异常信号通常可以表示为:

$$f(t) = \begin{cases} 1, t \geqslant \tau \\ 0, t < \tau \end{cases} \tag{4.12}$$

对于使用上式表示的均值阶跃型预期异常信号,参数 γ 表示均值发生的阶跃变化的幅度。图 4-4(a)给出了仿真实验中时间子序列 Y'_d 的正常模式窗口数据,图 4-4(b)给出了仿真实验中当过程发生均值阶跃时的时间子序列 Y'_d 的异常模式窗口数据,仿真实验中的窗口大小为 $l_w=28$,均值阶跃信号的幅度 $\gamma=2.5$,ARMA(1,1)模型参数 $\phi_1=0.6, \theta_1=-0.4$。

研究中 LS-SVM 模式识别器在不同的参数条件下进行训练,每种模式使用的训练样本容量为 20。当过程发生均值漂移型异常时,也可以将过程的异常模式称为"漂移模式(Shifted Pattern)"。

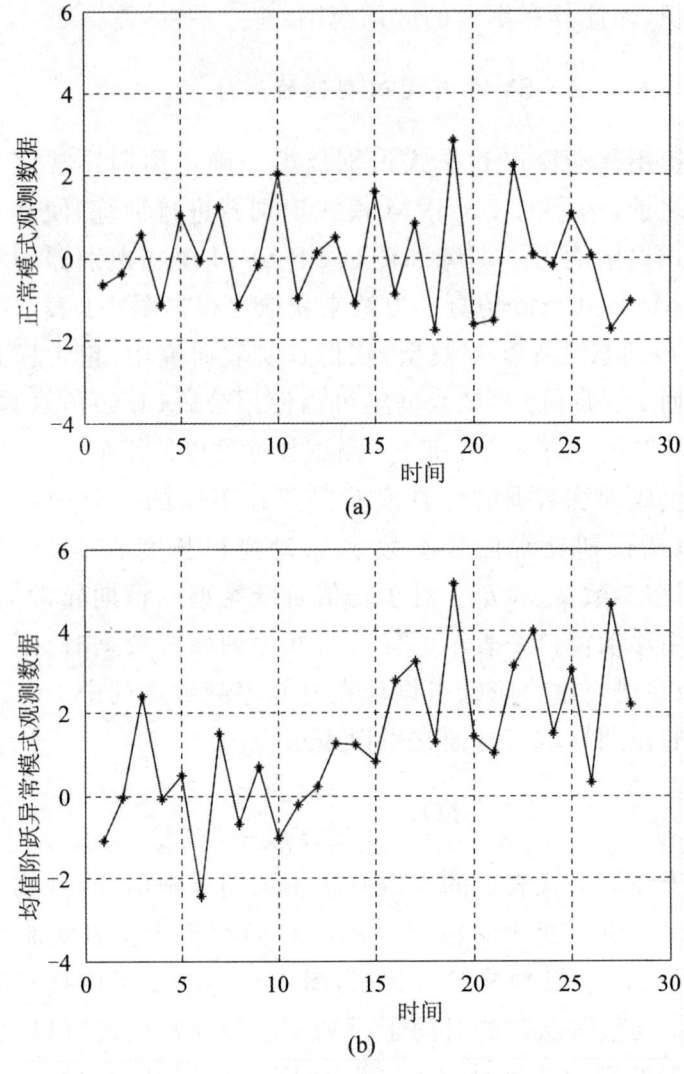

图 4-4

(a)一个窗口的正常模式训练数据；
(b)一个窗口的均值阶跃异常模式训练数据

4.3.4 模型性能评估

研究图 4-3 中模型的性能需要从两个方面对模型进行评估，首先需要评估 LS-SVM 模式识别器的性能；然后，在保证 LS-

第4章 基于Cuscore统计量的过程质量智能诊断

SVM模式识别器获得良好的模式识别能力的基础上，使用LS-SVM模式识别器确定Cuscore图的触发时刻，并使用ARL指标评价使用模式识别方法进行智能触发的Cuscore图对于预期异常信号的检测能力。

4.3.4.1 识别率性能评估

在使用LS-SVM模式识别技术分析时间序列的变点时，必须首先对LS-SVM模式识别器的性能进行测试和评估，保证LS-SVM模式识别器具有较高的模式识别率。为测试LS-SVM模式识别器的识别率性能，对应于生成正常/异常模式训练样本的参数和方法，生成正常/异常模式测试样本各1000个，测试样本的生成方法与训练样本的生成方法相同。仿真实验中涉及的可调节参数包括γ、l_w、ϕ_1和θ_1。对于LS-SVM识别器参数C和σ，使用交叉验证法进行优化选择，最后得到$C=10$，$\sigma=1$，该参数应用于所有实验，即，表4-3的实验结果都是在LS-SVM识别器参数$C=10$，$\sigma=1$下得到。对于完成了训练的LS-SVM识别器，使用测试样本数据对其性能进行测试和评估。LS-SVM识别器对于时间子序列的正常/异常模式识别能力使用识别率(Accuracy Rate, AR)来评估。在多种实验参数条件下，通过MATLAB仿真实验得到的LS-SVM模式识别器的识别率总结在表4-3中。

表4-3　LS-SVM识别器对于正常/异常时间序列窗口模式的识别率
（括号中为正常时间序列窗口模式的识别率）

γ	l_w	(a) $\phi_1=-0.8$ $\theta_1=0.2$	(b) $\phi_1=-0.6$ $\theta_1=0.4$	(c) $\phi_1=-0.4$ $\theta_1=0.6$	(d) $\phi_1=-0.2$ $\theta_1=0.8$	(e) $\phi_1=0.8$ $\theta_1=-0.2$	(f) $\phi_1=0.6$ $\theta_1=-0.4$	(g) $\phi_1=0.4$ $\theta_1=-0.6$	(h) $\phi_1=0.2$ $\theta_1=-0.8$
1	2	76.9 (42.7)	60.9 (60.1)	62.0 (57.9)	76.9 (42.7)	59.9 (55.0)	67.8 (55.8)	58.2 (56.5)	58.7 (48.1)
	6	60.5 (59.3)	62.3 (59.4)	65.4 (54.9)	56.2 (52.3)	77.1 (73.2)	76.3 (73.0)	69.9 (76.0)	67.1 (75.9)

续表

γ	l_w	(a) $\phi_1=-0.8$ $\theta_1=0.2$	(b) $\phi_1=-0.6$ $\theta_1=0.4$	(c) $\phi_1=-0.4$ $\theta_1=0.6$	(d) $\phi_1=-0.2$ $\theta_1=0.8$	(e) $\phi_1=0.8$ $\theta_1=-0.2$	(f) $\phi_1=0.6$ $\theta_1=-0.4$	(g) $\phi_1=0.4$ $\theta_1=-0.6$	(h) $\phi_1=0.2$ $\theta_1=-0.8$
	10	61.8 (53.7)	62.9 (62.2)	61.4 (47.7)	56.5 (51.4)	73.6 (87.4)	80.3 (73.3)	73.7 (83.1)	79.6 (70.2)
	14	60.1 (53.3)	61.5 (65.7)	55.2 (54.2)	58.6 (53.5)	77.3 (83.7)	80.6 (80.9)	76.4 (76.0)	75.7 (71.9)
	18	63.3 (56.0)	60.6 (63.7)	49.3 (52.1)	60.2 (52.9)	77.5 (73.7)	77.9 (78.9)	76.6 (75.7)	75.6 (67.6)
	22	62.1 (59.1)	69.3 (56.5)	51.2 (62.2)	56.7 (46.8)	83.2 (76.5)	81.3 (65.1)	80.6 (61.7)	79.0 (60.6)
	26	61.2 (51.0)	65.2 (50.8)	67.1 (47.4)	63.7 (37.4)	91.7 (64.7)	75.3 (72.2)	82.2 (61.9)	89.3 (35.9)
	30	80.2 (26.2)	86.2 (30.5)	73.8 (39.5)	85.4 (19.2)	96.1 (33.5)	95.5 (32.1)	95.4 (28.3)	96.5 (27.0)
	34	92.3 (19.2)	92.8 (11.4)	92.0 (12.5)	88.6 (14.7)	98.7 (18.9)	98.7 (17.9)	98.9 (15.6)	98.2 (13.1)
	38	98.8 (1.3)	95.2 (8.4)	98.5 (5.1)	96.3 (5.7)	99.6 (6.0)	99.8 (6.4)	99.9 (2.9)	99.7 (2.4)
	42	99.6 (1.0)	99.6 (1.9)	99.2 (1.1)	98.8 (0.5)	100 (1.4)	100 (1.0)	100 (1.8)	100 (0.3)
1.5	2	71.5 (61.9)	71.7 (78.6)	76.1 (54.8)	67.3 (61.6)	66.6 (62.6)	76.1 (73.1)	83.1 (61.0)	73.0 (52.2)
	6	71.8 (63.6)	66.8 (62.3)	59.2 (50.2)	57.0 (61.2)	92.4 (90.6)	87.6 (93.2)	84.8 (83.5)	96.1 (89.1)
	10	74.0 (73.1)	62.7 (77.6)	64.5 (60.0)	62.1 (63.6)	94.3 (92.5)	92.5 (93.1)	91.7 (93.5)	89.2 (84.2)
	14	70.1 (70.0)	76.0 (68.1)	58.1 (77.1)	67.3 (55.1)	94.9 (92.3)	95.6 (94.8)	92.4 (92.5)	86.9 (89.5)

第 4 章 基于 Cuscore 统计量的过程质量智能诊断

续表

γ	l_w	(a) $\phi_1=-0.8$ $\theta_1=0.2$	(b) $\phi_1=-0.6$ $\theta_1=0.4$	(c) $\phi_1=-0.4$ $\theta_1=0.6$	(d) $\phi_1=-0.2$ $\theta_1=0.8$	(e) $\phi_1=0.8$ $\theta_1=-0.2$	(f) $\phi_1=0.6$ $\theta_1=-0.4$	(g) $\phi_1=0.4$ $\theta_1=-0.6$	(h) $\phi_1=0.2$ $\theta_1=-0.8$
	18	73.0 (63.0)	63.8 (75.0)	70.3 (61.5)	61.2 (52.8)	96.0 (96.0)	96.0 (92.6)	95.4 (92.8)	92.0 (83.8)
	22	78.1 (73.0)	71.3 (54.3)	73.0 (61.1)	74.6 (50.1)	96.8 (86.4)	97.3 (88.2)	95.3 (87.5)	94.4 (74.5)
	26	85.9 (51.3)	81.9 (50.9)	77.2 (47.1)	67.8 (43.2)	99.6 (67.4)	99.3 (69.8)	98.0 (65.4)	98.8 (51.1)
	30	90.7 (31.6)	82.3 (44.6)	93.4 (22.6)	83.1 (22.6)	99.9 (41.3)	100 (32.4)	99/9 (39.3)	99.3 (28.5)
	34	95.7 (17.9)	95.1 (12.5)	92.0 (12.8)	94.8 (11.4)	100 (25.3)	100 (20.3)	100 (18.3)	99.9 (11.9)
	38	99.6 (4.8)	98.6 (5.1)	99.2 (3.9)	96.6 (4.7)	100 (8.7)	100 (4.0)	100 (4.2)	100 (3.2)
	42	99.9 (0.9)	99.2 (2.4)	99.6 (1.5)	98.7 (1.2)	100 (0.5)	100 (1.5)	100 (0.4)	100 (0.6)
2	2	80.0 (83.6)	84.4 (75.0)	78.7 (68.2)	74.4 (71.2)	83.1 (72.6)	74.7 (83.4)	79.3 (61.3)	75.5 (74.3)
	4	80.0 (74.4)	75.8 (69.2)	75.9 (60.1)	70.1 (68.9)	95.1 (94.7)	91.6 (94.0)	93.5 (96.9)	96.8 (90.2)
	6	80.4 (76.3)	76.9 (71.8)	68.0 (62.9)	78.5 (66.9)	95.6 (98.2)	95.6 (94.9)	93.3 (96.2)	93.9 (97.7)
	8	78.0 (73.5)	74.8 (71.1)	67.2 (67.2)	65.6 (77.0)	99.0 (98.0)	99.0 (98.8)	99.2 (93.2)	96.3 (95.3)
	10	82.4 (77.2)	80.5 (63.5)	65.7 (64.9)	63.5 (69.4)	99.2 (99.2)	99.5 (99.1)	98.9 (99.2)	98.5 (96.3)
	12	89.4 (77.0)	75.8 (79.6)	64.7 (69.6)	65.0 (47.9)	99.9 (99.0)	98.9 (99.5)	98.2 (98.5)	98.2 (97.1)

续表

γ	l_w	(a) $\phi_1=-0.8$ $\theta_1=0.2$	(b) $\phi_1=-0.6$ $\theta_1=0.4$	(c) $\phi_1=-0.4$ $\theta_1=0.6$	(d) $\phi_1=-0.2$ $\theta_1=0.8$	(e) $\phi_1=0.8$ $\theta_1=-0.2$	(f) $\phi_1=0.6$ $\theta_1=-0.4$	(g) $\phi_1=0.4$ $\theta_1=-0.6$	(h) $\phi_1=0.2$ $\theta_1=-0.8$
	14	83.6 (85.5)	80.8 (75.0)	78.7 (71.1)	69.5 (63.7)	99.5 (98.9)	99.6 (99.4)	99.4 (97.8)	99 (97.8)
	16	81.1 (89.1)	78.9 (84.9)	76.1 (76.5)	64.8 (59.9)	99.8 (98.8)	99.4 (98.9)	99.1 (97.2)	96.8 (96.1)
	18	87.0 (82.5)	79.5 (80.2)	73.8 (77.7)	77.2 (55.9)	99.9 (96.5)	99.2 (97.2)	99.5 (99.2)	98.2 (93.2)
	20	90.2 (78.1)	86.4 (79.1)	84.1 (78.2)	83.8 (49.2)	100 (95.8)	100 (94.9)	99.6 (92.4)	99.3 (87.9)
	22	81.8 (86.0)	87.3 (69.4)	85.6 (69.7)	82.4 (36.0)	100 (92.9)	99.9 (88.1)	99.8 (82.6)	99.2 (75.9)
	24	92.9 (74.8)	83.5 (73.6)	88.0 (66.7)	79.1 (51.9)	100 (80.1)	100 (80.6)	100 (77.6)	99.9 (58.5)
	26	96.0 (62.7)	87.3 (68.9)	86.7 (59.4)	68.7 (41.1)	100 (65.9)	100 (62.2)	100 (59.2)	100 (51.2)
	28	93.3 (61.4)	97.8 (39.3)	84.5 (33.9)	81.1 (39.0)	100 (49.4)	100 (57.3)	100 (38.9)	100 (37.4)
	30	93.6 (45.8)	96.6 (41.2)	91.6 (25.3)	89.2 (28.4)	100 (43.3)	100 (44.3)	100 (40.1)	100 (20.1)
	32	98.1 (29.1)	98.4 (18.7)	97.2 (19.4)	92.8 (16.9)	100 (25.7)	100 (25.4)	100 (28.8)	100 (19.6)
	34	98.7 (18.0)	95.9 (21.8)	97.3 (15.5)	94.8 (11.8)	100 (14.5)	100 (13.8)	100 (14.1)	100 (10.9)
	36	99.3 (10.3)	99.8 (6.1)	98.5 (7.2)	97.0 (5.4)	100 (6.6)	100 (11.9)	100 (7.5)	100 (10.1)
	38	99.8 (9.3)	99.9 (5.9)	98.9 (5.4)	98.7 (3.9)	100 (4.5)	100 (3.7)	100 (5.3)	100 (3.8)

第 4 章 基于 Cuscore 统计量的过程质量智能诊断

续表

γ	L_w	(a) $\phi_1=-0.8$ $\theta_1=0.2$	(b) $\phi_1=-0.6$ $\theta_1=0.4$	(c) $\phi_1=-0.4$ $\theta_1=0.6$	(d) $\phi_1=-0.2$ $\theta_1=0.8$	(e) $\phi_1=0.8$ $\theta_1=-0.2$	(f) $\phi_1=0.6$ $\theta_1=-0.4$	(g) $\phi_1=0.4$ $\theta_1=-0.6$	(h) $\phi_1=0.2$ $\theta_1=-0.8$
	40	99.9 (2.1)	99.9 (3.4)	99.9 (1.7)	98.7 (2.0)	100 (2.3)	100 (2.8)	100 (2.9)	100 (2.5)
	42	100 (2.2)	100 (1.3)	99.8 (1.2)	100 (0.7)	100 (0.9)	100 (1.3)	100 (1.8)	100 (0.6)
2.5	2	82.2 (88.8)	89.5 (80.2)	83.6 (92.0)	83.8 (86.1)	86.7 (86.4)	86.2 (84.4)	83.9 (83)	91.7 (78.0)
	6	82.8 (82.7)	82.6 (76.5)	72.3 (76.8)	84.5 (74.2)	99.8 (99.7)	99.9 (99.6)	99.7 (99.6)	98.9 (98.4)
	10	87.4 (84.4)	90.8 (78.5)	81.6 (82.2)	81.9 (73.2)	99.9 (100)	99.9 (100)	99.9 (99.1)	99.5 (98.7)
	14	92.0 (92.8)	83.2 (86.3)	81.4 (87.9)	68.9 (71.3)	100 (99.6)	100 (99.7)	99.8 (99.9)	99.3 (99.2)
	18	92.2 (90.0)	85.8 (79.5)	86.6 (85.0)	75.9 (59.0)	100 (98.0)	100 (97.1)	100 (96.6)	99.7 (95.1)
	22	96.1 (82.8)	89.6 (84.6)	87.2 (75.0)	77.7 (66.3)	100 (92.3)	100 (88.5)	100 (86.9)	100 (82.0)
	26	96.6 (67.1)	94.1 (70.0)	85.4 (55.5)	78.2 (46.5)	100 (64.2)	100 (50.4)	100 (67.0)	100 (51.7)
	30	99.5 (38.6)	99.0 (38.1)	92.4 (33.5)	93.8 (24.0)	100 (43.1)	100 (42.3)	100 (39.7)	100 (27.3)
	34	100 (19.9)	98.9 (17.3)	98.7 (16.3)	98.1 (11.5)	100 (18.3)	100 (17.4)	100 (17.4)	100 (12.7)
	38	100 (3.8)	100 (7.8)	99.6 (4.5)	99.2 (3.1)	100 (5.1)	100 (3.2)	100 (5.0)	100 (4.8)
	42	100 (3.0)	100 (2.4)	99.9 (1.2)	99.8 (0.9)	100 (1.7)	100 (1.2)	100 (1.8)	100 (1.1)

续表

γ	l_w	(a) $\phi_1=-0.8$ $\theta_1=0.2$	(b) $\phi_1=-0.6$ $\theta_1=0.4$	(c) $\phi_1=-0.4$ $\theta_1=0.6$	(d) $\phi_1=-0.2$ $\theta_1=0.8$	(e) $\phi_1=0.8$ $\theta_1=-0.2$	(f) $\phi_1=0.6$ $\theta_1=-0.4$	(g) $\phi_1=0.4$ $\theta_1=-0.6$	(h) $\phi_1=0.2$ $\theta_1=-0.8$
3	2	89.1 (86.1)	93.2 (89.7)	91.0 (93.5)	89.7 (90.8)	91.0 (84.5)	90.6 (87.3)	87.4 (87.6)	96.2 (87.2)
	6	84.2 (90.1)	88.5 (84.0)	91.0 (85.4)	84.2 (82.3)	100 (99.6)	100 (99.8)	99.9 (100)	100 (99.9)
	10	91.1 (94.5)	91.8 (81.8)	90.4 (83.5)	81.0 (81.7)	100 (99.9)	100 (99.9)	99.9 (99.9)	99.9 (98.6)
	14	94.3 (96.6)	89.4 (93.1)	82.6 (81.1)	80.3 (84.8)	100 (99.9)	100 (99.9)	100 (99.8)	99.8 (98.0)
	18	96.0 (96.0)	93.0 (88.2)	92.3 (86.8)	87.0 (77.8)	100 (98.2)	100 (98.1)	100 (94.5)	99.8 (93.2)
	22	94.6 (88.8)	96.3 (86.1)	93.3 (79.4)	89.6 (70.4)	100 (89.5)	100 (91.2)	100 (84.6)	100 (79.0)
	26	99.3 (69.2)	98.7 (64.7)	96.9 (60.9)	91.2 (53.8)	100 (65.8)	100 (66.8)	100 (62.7)	100 (55.2)
	30	98.1 (42.9)	99.9 (45.4)	97.0 (39.1)	96.0 (21.1)	100 (42.6)	100 (50.1)	100 (37.8)	100 (29.6)
	34	100 (15.6)	100 (20.1)	99.5 (16.7)	97.6 (7.7)	100 (16.1)	100 (13.7)	100 (13.1)	100 (12.0)
	38	100 (3.5)	100 (7.2)	99.8 (8.5)	99.4 (2.1)	100 (5.3)	100 (4.7)	100 (4.7)	100 (2.8)
	42	100 (1.6)	100 (1.7)	100 (2.1)	99.8 (0.9)	100 (0.5)	100 (1.1)	100 (1.3)	100 (1.0)

第4章 基于 Cuscore 统计量的过程质量智能诊断

根据表 4-3 的实验结果,可以得到下面的结论。

(1) 对于幅度值大的均值漂移信号,LS-SVM 识别器具有更好的识别率性能。这表明图 4-3 中的智能触发 Cuscore 图的方法适用于检测过程中存在的幅度值较大的预期异常信号。

(2) LS-SVM 识别器对于正常/异常时间序列窗口模式的识别率受 ARMA(1,1) 参数 ϕ_1 和 θ_1 的影响较大。这是由于 ARMA(1,1) 时间序列的方差随着参数 ϕ_1 和 θ_1 的不同而产生较大变化,LS-SVM 识别器对于方差较大的 ARMA(1,1) 时间序列模式具有更好的模式识别性能。

(3) 为了使得 LS-SVM 识别器对于正常/异常时间序列窗口模式都具有优良的识别率性能,需要合理选择移动窗口的长度 l_w,我们通过仿真实验的方法来确定 l_w 的合理取值,该方法的原理可以通过图 4-5 来说明。图 4-5 是通过仿真实验得到的表 4-3 中(f)列的数据(当 $\gamma=2$ 时)表示的识别率与窗口长度 l_w 之间的关系图。从图 4-5 可知,当窗口长度为 6,8,10,12,14,16,18 和 20 时,LS-SVM 识别器对于正常模式和异常模式的识别率都很高。但是从质量成本的角度,在识别率满足要求的情况下,选择较小的时间序列窗口有利于节约质量成本。另外,从图 4-5 可以观察到,当移动窗口长度很大时,异常模式的识别率接近于 100%,而正常模式的识别率下降到几乎为 0。根据由实验得到的表 4-3 和图 4-5,可以合理确定移动窗口长度的 l_w。

4.3.4.2 ARL 性能评估

对于图 4-3 中使用 LS-SVM 模式识别技术进行触发的 Cuscore 图,其对于预期异常信号的检测性能需要加以评估。我们在研究中使用 Monte-Carlo 仿真的方法,对过程中出现均值漂移预期异常信号时智能触发的 Cuscore 图和标准 Cuscore 图的失控 ARL 指标进行研究和评估。均值漂移型的信号是过程中最为常见的预期异常信号。为评估和比较智能触发的 Cuscore 图和标准 Cuscore 图监控过程异常时的失控 ARL,按照下面的步骤进行了 MATLAB 仿真实验。

(1)使用图 4-1 中的模型和 4.2.1 节中的方法来计算这两种 Cuscore 图的控制限。

(2)使用方程(4.7)产生时间序列,并且使得公式(4.12)中的均值漂移预期异常信号发生阶跃的时间 τ 在[1,40]的范围内满足均匀分布,即 $\tau \sim U(1,40)$。

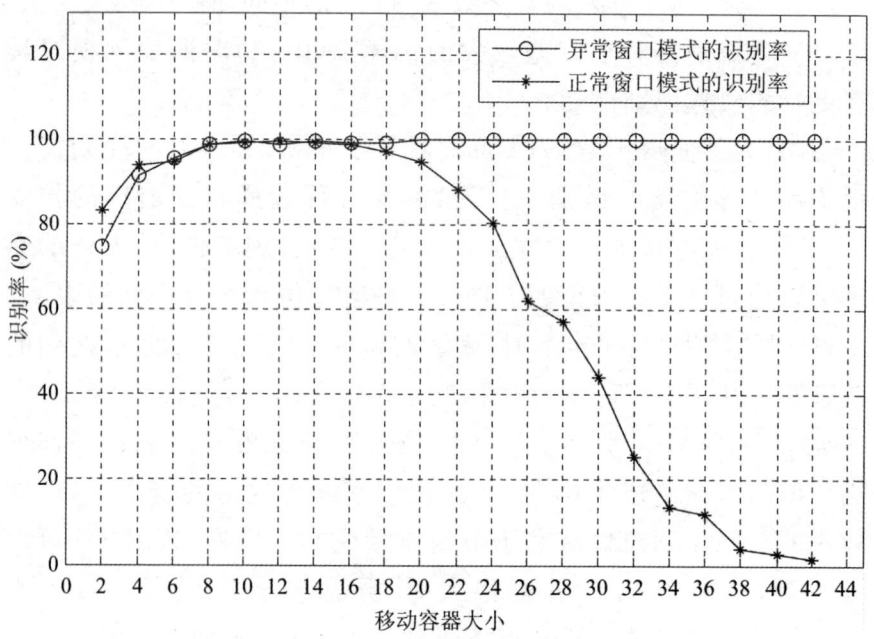

图 4-5 识别率和移动窗口长度关系曲线图

(3)计算智能触发的 Cuscore 图和标准 Cuscore 图的 Cuscore 统计量,并找到它们各自的一个失控链长。

(4)重复步骤(2)、(3)1000 次,可以计算得到智能触发的 Cuscore 图和标准 Cuscore 图的失控 ARL。

上面的计算智能触发的 Cuscore 图的 ARL 的实验,必须合理选择参数 γ、l_w、ϕ_1 和 θ_1,保证 LS-SVM 识别器具有较高的识别率,否则实验结果没有意义。表 4-4 给出了仿真实验在不同参数取值时的结果。

第 4 章 基于 Cuscore 统计量的过程质量智能诊断

表 4-4 智能触发的 Cuscore 图和标准 Cuscore 图的 ARL 性能

(括号中的数据为标准 Cuscore 图的 ARL)

γ	(a) $\phi_1=-0.8$ $\theta_1=0.2$	(b) $\phi_1=-0.6$ $\theta_1=0.4$	(c) $\phi_1=-0.4$ $\theta_1=0.6$	(d) $\phi_1=-0.2$ $\theta_1=0.8$	(e) $\phi_1=0.8$ $\theta_1=-0.2$	(f) $\phi_1=0.6$ $\theta_1=-0.4$	(g) $\phi_1=0.4$ $\theta_1=-0.6$	(h) $\phi_1=0.2$ $\theta_1=-0.8$
				$\tau \sim U(1,40), l_w=16$				
1.0	102.8 (104.5)	103.8 (109.1)	107.3 (108.4)	124.5 (125.5)	21.3 (21.0)	33.5 (34.1)	45.1 (47.4)	37.6 (39.8)
1.2	68.9 (76.6)	71.6 (83.8)	70.0 (78.5)	92.9 (100.7)	11.1 (15.6)	18.9 (26.6)	29.6 (38.7)	30.7 (38.4)
1.4	46.8 (62.3)	57.5 (70.0)	60.8 (72.8)	73.8 (85.5)	7.6 (14.8)	11.6 (21.2)	17.1 (28.3)	16.9 (28.2)
1.6	39.4 (57.1)	43.9 (58.8)	56.0 (72.6)	58.9 (72.2)	6.9 (14.5)	9.5 (20.0)	12.5 (25.4)	13.2 (27.3)
1.8	33.2 (49.4)	38.8 (54.1)	46.9 (62.2)	60.7 (77.0)	5.7 (13.6)	7.0 (20.4)	10.1 (23.3)	11.7 (26.3)
2.0	27.9 (43.0)	36.4 (51.8)	35.2 (53.4)	51.7 (66.2)	5.4 (14.7)	6.1 (16.8)	8.6 (22.7)	9.9 (24.9)
2.2	27.1 (43.2)	25.5 (42.8)	34.9 (53.9)	42.3 (58.6)	4.6 (14.2)	6.0 (17.6)	7.4 (23.5)	8.0 (23.3)
2.4	27.9 (45.0)	28.1 (46.9)	27.9 (7.4)	34.0 (49.7)	4.8 (14.1)	5.8 (19.6)	6.1 (20.6)	7.6 (23.6)
2.6	20.3 (36.7)	25.2 (42.4)	27.6 (46.8)	31.6 (50.2)	4.3 (13.8)	4.9 (17.3)	6.0 (21.3)	6.4 (22.1)

115

续表

γ	\multicolumn{8}{c}{$\tau \sim U(1,40), l_w=16$}
	(a) $\phi_1=-0.8$ $\theta_1=0.2$	(b) $\phi_1=-0.6$ $\theta_1=0.4$	(c) $\phi_1=-0.4$ $\theta_1=0.6$	(d) $\phi_1=-0.2$ $\theta_1=0.8$	(e) $\phi_1=0.8$ $\theta_1=-0.2$	(f) $\phi_1=0.6$ $\theta_1=-0.4$	(g) $\phi_1=0.4$ $\theta_1=-0.6$	(h) $\phi_1=0.2$ $\theta_1=-0.8$
2.8	20.9 (38.3)	23.6 (41.7)	25.9 (44.6)	32.4 (50.0)	4.0 (13.8)	4.5 (18.1)	5.0 (19.8)	5.7 (22.0)
3.0	20.6 (36.7)	19.0 (35.3)	22.2 (39.6)	24.9 (42.2)	4.0 (14.9)	4.4 (18.5)	5.2 (21.5)	5.6 (21.6)

从表 4-4 可以得到如下实验结论。

(1)在实验选定的各种参数条件下,智能触发的 Cuscore 图的 ARL 普遍小于标准 Cuscore 图的 ARL。这表明智能触发的 Cuscore 图对于均值漂移异常信号的检测能力好于标准 Cuscore 图。

(2)在 ARMA(1,1)参数 ϕ_1、θ_1 保持不变时,智能触发的 Cuscore 图和标准 Cuscore 图的 ARL 都随着幅度参数 γ 的增大而减小。

(3)在幅度参数 γ 保持不变时,ARMA(1,1)模型参数 ϕ_1、θ_1 参数对智能触发的 Cuscore 图和标准 Cuscore 图的 ARL 性能有重要影响。根据表 4-4,对于列(e)、(f)、(g)、(h)的 ϕ_1 和 θ_1 取值,其失控 ARL 值相对较小;对于列(a)、(b)、(c)、(d)的 ϕ_1 和 θ_1 取值,其失控 ARL 值相对较大。

4.3.5 仿真算例

下面的仿真算例用于说明如何使用智能触发的 Cuscore 图和标准 Cuscore 图来监控过程异常。按照图 4-1 中介绍过程模型,算例中假定过程的初始 30 个过程观测数据处于受控状态,从第 31 个数据开始,过程中发生了幅值为 $\gamma=2$ 的均值漂移异常情况,ARMA(1,1)模型参数选取 $\phi_1=-0.2$,$\theta_1=0.8$。由于均值漂

第 4 章 基于 Cuscore 统计量的过程质量智能诊断

移信号是单向变化的,现使用 Q^+ 统计量分别绘制智能触发的 Cuscore 图和标准 Cuscore 图对该过程进行异常诊断。仿真研究中利用过程受控时 ARL 为 370 的概念,通过仿真来计算智能触发的 Cuscore 图和标准 Cuscore 图的控制限,得到 Q^+ 统计量的控制限为 UCL=18.44;使用窗口长度 $l_w = 16$ 的模式识别器来估计均值漂移出现的时刻 τ,最后得到智能触发 Cuscore 图如图 4-6 所示,标准 Cuscore 图如图 4-7 所示。

根据图 4-6,智能 Cuscore 图在第 63 个仿真数据处首次发出报警,但 LS-SVM 模式识别器估计出均值漂移发生的时间为 $\hat{\tau}=31$,智能 Cuscore 图在该时刻进行触发,因此智能触发的 Cuscore 图对于均值漂移信号的报警延迟为 63−31=32 个数据单元。根据图 4-7,使用标准 Cuscore 图监控过程时,标准 Cuscore 图直到第 62 个仿真数据处才发出异常报警,由于标准 Cuscore 图在第一个仿真数据处进行触发,因此标准 Cuscore 图对于均值漂移信号

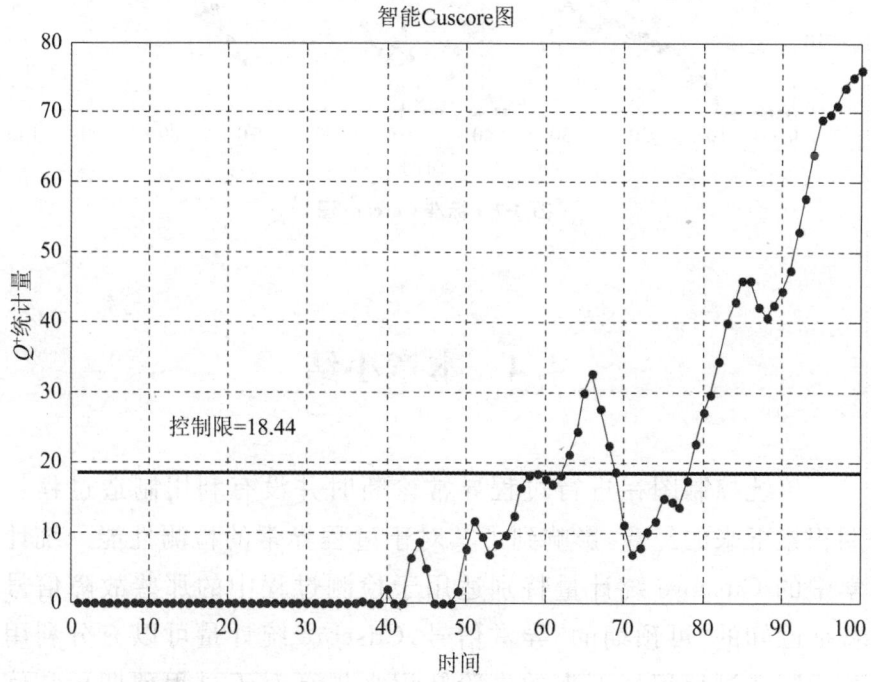

图 4-6 智能触发的 Cuscore 图

的报警延迟为 62 个数据单元。上面的仿真算例虽然仅针对均值漂移异常信号进行了诊断研究,并没有涉及更多的其他类型的异常信号,但仍然可以看出智能触发的 Cuscore 图比标准的 Cuscore 图具有更好的过程异常诊断能力。

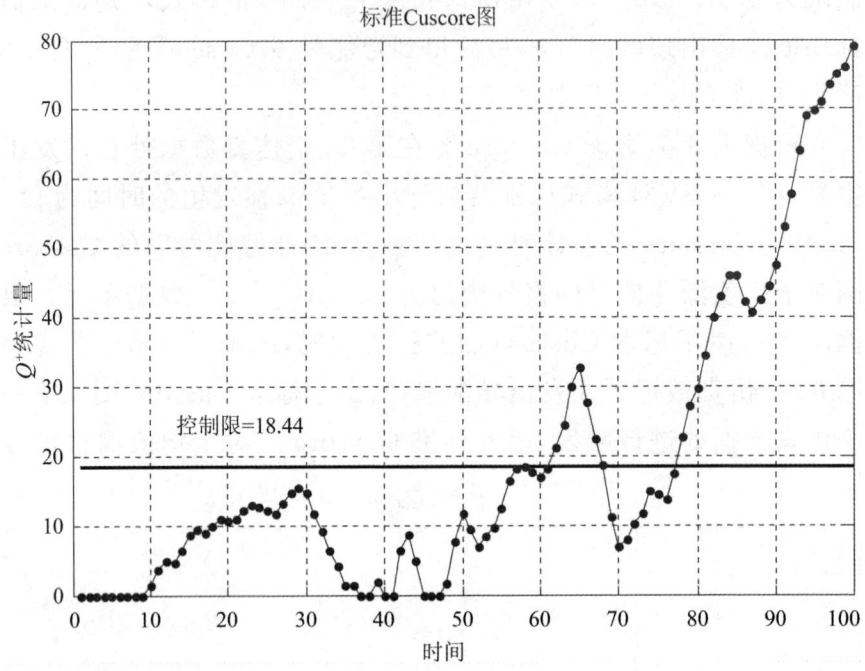

图 4-7　标准 Cuscore 图

4.4　本章小结

传统控制图在进行过程异常诊断时并没有利用制造过程长期积累下来的经验,影响到了其对于过程异常的检测性能。统计学中的 Cuscore 统计量特别适用于检测过程中的那些故障信号特征已知的"可预期的"异常信号,Cuscore 统计量可以充分利用关于制造过程积累下来的先验知识来提高对于过程预期异常信号的诊断效率。

第 4 章 基于 Cuscore 统计量的过程质量智能诊断

本章介绍了使用 Cuscore 统计量监控过程异常的基本原理和特点，在过程数据存在自相关的情况下建立了单工序过程的 ARMA 模型；研究了 Cuscore 图对于非线性二次预期信号的诊断性能；提出使用移动窗口法和 LS-SVM 模式识别技术来进行变点检测，以解决标准 Cuscore 图中存在的"失配"问题，并通过仿真研究对提出的有关模型和技术方法的性能进行了评估。过程质量诊断需要根据制造过程的具体情况来确定适合的诊断技术，本章对于如何使用 Cuscore 统计量进行过程质量监控等问题进行了研究与讨论。

第 5 章 多元过程质量智能诊断与异常变量识别

实际生产中的制造过程大多包含多个质量特性变量,多元过程的质量诊断与异常变量分离问题是质量管理领域最具挑战性的研究方向之一。本章针对使用传统统计学方法诊断多元过程均值矢量和协方差矩阵异常时存在的只能判断过程的总体状态,无法具体识别和定位异常变量的问题,分别提出使用模式识别技术对多元过程均值矢量和协方差矩阵异常进行诊断的智能模型,解决了多元过程质量诊断中的异常变量的识别与定位问题。通过仿真分析和实例分析,验证了所提出方法的可行性和有效性。

5.1 多元过程质量诊断问题

最早发明的休哈特质量控制图是针对产品的单一质量特性进行监控和诊断的,而在实际的过程质量控制中,多数情况下需要同时监控多个质量特性变量。一般人们将具有多个质量特性的过程称为多元过程或者多变量过程。对于多变量过程质量诊断问题,人们自然容易想到对多变量过程的每一个变量分别采用单变量休哈特控制图进行过程质量监控与诊断。另外,传统的SPC技术建立在过程观测数据满足独立同分布的假设之上,然而,实际的很多制造过程并不满足这一假设条件,比如,化工、制药等连续型生产过程的质量特性变量之间往往具有相关性。对于多元过程,如果简单地将单变量休哈特控制图分别应用于多元

第5章 多元过程质量智能诊断与异常变量识别

过程的各个质量特性变量进行监控,当每个质量特性的观测值都控制在其控制界限内时就认为过程正常,这种做法往往会发生过高误判现象,尤其是当多元过程的质量特性变量之间存在相关性时,误判现象将更加严重[1]。这种方法处理过程观测数据的实质是,在多元过程情况下,如果对于相关过程数据采用各变量指标割裂开来分别进行控制的方法,反映质量特性指标之间相关关系的协方差系数就一个也没有加以控制,将多元质量控制问题简单化处理时,只控制了部分参数,导致对过程状态过高的错误判断(误判和漏判),引起对过程的过调整和欠调整,给企业带来经济损失。针对这种问题,需要质量工作者研究和发展多元统计过程控制(Multivariate Statistical Process Control,MSPC)理论。也就是说,不能将传统的单变量休哈特控制图简单地推广到多变量的情况,不能使用对各个变量的分别监控来代替对整个过程的全部变量进行联合控制,否则就会发生过高误判现象。此时需要研究使用多元质量控制方法。

Hotelling最早注意到将单变量控制图简单推广应用于多变量过程的缺陷,并于1947年提出多元质量控制的概念[161],关于多元质量控制的许多其他概念都是在此基础上发展起来的。多元统计过程控制与诊断理论和一元质量控制与诊断理论相类似,只不过多元过程质量诊断理论多数是建立在多元过程满足多元正态分布的假设之上,通过建立某个检验统计量对过程进行质量异常诊断,多元SPC控制图的实质仍然是一个假设检验问题。

人工智能是工程技术领域中引人注目的技术成果,并且处于不断的发展中。质量管理与诊断属于交叉学科,将人工智能技术引入多元过程质量诊断,必然会不断产生新的基于人工智能技术的质量诊断有效方法,提高制造业质量诊断与管理水平。本章下面的研究重点是将人工智能技术应用于多元过程质量诊断,建立基于人工智能技术的质量诊断模型,实现对过程异常变量的辨识。

5.2 多元统计过程控制图

多元质量异常诊断技术的研究和应用是制造过程发展到一定阶段的必然要求,这一问题在普遍采用先进制造技术和全面质量管理的今天尤其显得重要。比如,在化工制造和半导体制造行业,通常需要监控数百个质量特性变量,同时,测试计量技术、计算机技术以及数据库等技术的发展,为采集和处理这些质量数据提供了更加先进的技术手段。

多元统计过程控制(MSPC)技术是一元统计过程控制技术的有效发展。使用统计学方法监控多元过程时,通常需要假设多元过程变量满足多元正态的条件,并在此前提下构造多元过程的总体检验统计量以便对多元过程的正常/异常状态进行判断。对满足一元正态分布的过程,其均值与方差这两个参数是互相独立的,故控制正态分布就需要对均值与方差分别应用相应的控制图进行控制。同理,在多元正态分布条件下,过程质量特性变量的均值向量与多元协方差矩阵这两个参数也是互相独立的,故诊断与控制服从多元正态分布的过程状态也需要分别使用相应的控制图对均值向量与多元协方差矩阵分别进行异常诊断与控制,由此对于多元过程的诊断与监控使用了两类多元控制图:多元均值控制图和多元方差控制图。这些基于多元统计学的控制图目前仍然是研究多元过程质量诊断的重要工具,在实际应用中获得了良好效果,其他的多元过程质量诊断技术的深入研究多数以此为基础而展开,下面分别对这两类 MSPC 控制图进行介绍。

5.2.1 多元均值控制图

目前在多元质量诊断领域里的大多数研究都是针对过程的均值矢量异常现象而进行。针对过程均值矢量异常进行诊断的具有代表性的多元均值控制图包括 Hotelling T^2 图、MCUSUM 图、MEWMA 图。

5.2.1.1 Hotelling T^2 图

对于一个具有 P 个质量特性变量的 P 维受控过程,它的 n 个采样样本可以表示为 $\{X_1, X_2, \cdots X_n\}$, $i = 1, 2, \cdots, n$,则 $X_i = (X_{i1}, X_{i2}, \cdots, X_{iP})^{\mathrm{T}}$ 表示多元过程在采样时刻 i 的第 i 个 P 维质量特性矢量观测数据。X_{ij} 表示第 j 个质量特性变量的第 i 个观测值。统计学研究中一般要求 $n > P$。使用 Hotelling T^2 图监控过程时,要求过程已经处于初始受控状态,X_i 是独立的服从均值矢量为 μ、协方差矩阵为 Σ 的变量,即 $X_i \sim N(\mu, \Sigma)$。对于实际的多元制造过程,μ 和 Σ 常常是未知的,此时可以使用受控过程的 n 个观测样本分别对其进行估计,得到受控过程的样本均值矢量 $\overline{X}$ 和样本协方差矩阵 S 分别为 $\overline{X} = (\overline{X}_1, \overline{X}_2, \cdots, \overline{X}_P)^{\mathrm{T}}$ 和 $S = \frac{1}{n-1} \sum_{i=1}^{n} (X_i - \overline{X})(X_i - \overline{X})^{\mathrm{T}}$,其中,$\overline{X}_j = \frac{1}{n} \sum_{i=1}^{n} X_{ij}$,那么,对于过程观测数据 X_i,可以使用下面的公式来构造它的 T^2 检验统计量:

$$T_i^2 = (X_i - \overline{X})^{\mathrm{T}} S^{-1} (X_i - \overline{X}) \tag{5.1}$$

由于 $T_i^2 \dfrac{n(n-P)}{P(n^2-1)}$ 服从自由度为 P 和 $(n-P)$ 的 F 分布,那么,在 Hotelling T^2 控制图理论中,对于给定的置信水平 α,观测数据 X_i 的 T^2 检验统计量的上下控制限分别使用下面的式子进行计算:

$$UCL_{T^2} = \frac{P(n^2-1)}{n(n-P)} F_{\alpha}(P, n-P) \tag{5.2}$$

$$LCL_{T^2} = 0 \tag{5.3}$$

特别地,当过程的参数(μ 和 Σ)已知时,T^2 统计量就是 χ^2 统计量;相应地,控制图的上限 UCL 变成为 $\chi^2_{\alpha,P}$。当观测数据 X_i 的 T^2 统计量超过了控制限,Hotelling T^2 控制图就发出报警信号,报告过程的异常状态。观测数据 X_i 的 T^2 检验统计量实际上表示了观测数据 X_i 到过程目标值(过程的稳态均值矢量)之间的距离。同其他的统计过程控制技术一样,使用 T^2 统计量进行过程异常诊断,同样存在第一类错误 I(α) 和第二类错误 II(β) 的问题。特别地,当过程中只有两个质量特性变量时,可以证明,T^2 为以这两个变量为自变量的椭圆方程,而 T^2 统计量的置信区间边界即为椭圆。

T^2 统计量经常以累计得分的方式加以使用,发展成为 COT (CUSUM of T) 控制图。对于大于零的正的常数 $k>0$,以迭代的方式定义累计得分统计量 s_i:

$$s_i = \max(0, s_{i-1} + T_i - k) \tag{5.4}$$

其中,$i=1,2,\cdots$。当 s_i 的值超过了一定的限度值 h,即当 $s_i > h$,COT 控制图发出过程异常报警信号。h 为达到一定的受控平均链长(In-Control ARL)而设定的控制限。受控平均链长的经验取值一般为 370 或 500,也可以根据对于过程的实际质量控制标准而设定。

5.2.1.2 MCUSUM 图

Crosier 提出的 MCUSUM 图由单变量 CUSUM 图发展而来。使用与 Hotelling T^2 控制图相同的过程数学模型,对于一个特定的常数 k,MCUSUM 统计量 S_i 以迭代的方式加以定义:

$$S_i = \begin{cases} 0, & \text{当 } C_n < k \\ (S_{i-1}+X_i-\mu)(1-k/C_i), & \text{其他} \end{cases} \tag{5.5}$$

其中,$C_i = [(S_{i-1}+X_i-\mu)^T \Sigma^{-1}(S_{i-1}+X_i-\mu)]^{1/2}$,$i=1,2,\cdots$。

对于某个特定的常数 h,当 $Y_i = (S_i^T \Sigma^{-1} S_i)^{1/2} > h$,MCUSUM 控制图发出过程均值矢量异常报警信号。同样地,h 为达到一定的受控平均链长(In-Control ARL)而设定的控制限。

5.2.1.3 MEWMA 图

Lowry 和 Woodall 等提出的 MEWMA 控制图也是由单变量 EWMA 图发展而来。使用与 Hotelling T^2 控制图相同的过程数学模型,MEWMA 统计量(矢量)可以以迭代的方式定义为:

$$Z_i = R(X_i - \mu) + (I-R)Z_{i-1} \tag{5.6}$$

其中，$i=1,2,\cdots$，R 为加权对角矩阵，$R = \text{diag}(r_1, r_2, \cdots, r_P)$，$0 \leqslant r_j \leqslant 1, j=1,2,\cdots,P$，$P$ 为多元过程质量特性变量的维数。在无任何关于过程的先验知识的情况下，加权系数一般取相同的值 $r_1 = r_2 = \cdots = r_P = r$。

对于某个特定的常数 h，当 $Y_i = (Z_i^T \Sigma^{-1} Z_i)^{1/2} > h$，MEWMA 控制图发出过程均值矢量异常报警信号。同样地，h 为达到一定的受控平均链长(In-Control ARL)而设定的控制限。

5.2.2 多元方差控制图

上面在讨论 Hotelling T^2 图、MEWMA 图、MCUSUM 图等多元均值 SPC 控制图时，需要假定这 P 个质量特性的协方差矩阵保持统计稳定状态，否则这些控制图将变得没有意义。事实上，多元过程的均值矢量独立于多元过程的协方差矩阵，它们需要分别加以控制。典型的监控多元过程散度变化的控制图为 Alt 提出的样本广义方差 $|S|$ 图，其对多元过程的异常具有很高检测与诊断能力。

统计学中，对于一个具有 P 个质量特性变量的多元过程，其散度可以使用 $P \times P$ 的协方差矩阵 Σ 来描述，并且将稳定受控过程的协方差矩阵记为 Σ_0。对于子组容量为 $n(n>P)$ 的一组样本 $X_1, X_2, \cdots, X_n$，其广义方差 $|S|$ 作为一个检验统计量，可以使用下面的公式得到，

$$|S| = \left| \frac{1}{n-1} \sum_{i=1}^{n} (X_i - \bar{X})(X_i - \bar{X})^T \right| \tag{5.7}$$

其中，$\bar{X} = \frac{1}{n}\sum_{i=1}^{n} X_i$。设 $E(|S|)$ 和 $V(|S|)$ 分别表示 $|S|$ 的均值和方差，根据控制图原理，$|S|$ 的大部分概率分布落在区间 $E(|S|) \pm 3\sqrt{V(|S|)}$ 中。可以证明，$|S|$ 图的上控制限、中心线、下控制限分别为：

$$UCL = |\Sigma_0|(b_1 + 3b_2^{1/2}) \qquad (5.8)$$

$$CL = b_1|\Sigma_0| \qquad (5.9)$$

$$LCL = \text{Max}\{0, |\Sigma_0|(b_1 - 3b_2^{1/2})\} \qquad (5.10)$$

系数 b_1 和 b_2 与过程质量特性变量的个数 P 有关,通过下面的式子计算得到:

$$b_1 = (n-1)^{-p} \prod_{i=1}^{p}(n-i) \qquad (5.11)$$

$$b_2 = (n-1)^{-2p} \prod_{i=1}^{p}(n-i)\left[\prod_{j=1}^{p}(n-j+2) - \prod_{j=1}^{p}(n-j)\right]$$

$$(5.12)$$

实际应用时,Σ_0 可以由样本数据估计得到。令 S_j 表示受控过程的第 j 个样本的协方差矩阵,$j=1,2,\cdots,m$,$\overline{S}$ 为 S_j 的均值,$|\overline{S}|$ 为 $\overline{S}$ 的行列式,则 $|\Sigma_0|$ 的估计值为 $|\overline{S}|/b_1$。

5.3　多元过程均值异常诊断与变量识别的智能诊断模型

在传统的 MSPC 技术的基础上,将包括数据挖掘、人工智能等多种计算机技术应用于过程异常诊断,丰富质量诊断工具并且提高过程诊断的效能,是质量工程领域具有发展前景的研究课题。

5.3.1　多元过程均值矢量异常的智能诊断模型

为了实现对多元过程均值矢量进行异常诊断并且实现异常变量的辨识与定位,我们首先假定过程中的异常因素仅仅使过程均值矢量发生异常变化,不会影响到过程的协方差矩阵和变量间的相关系数,在此前提下,我们提出使用 Hotelling T^2 控制图和 LS-SVM 模式识别技术的多元过程异常诊断模型,模型的结构如

第 5 章 多元过程质量智能诊断与异常变量识别

图 5-1 所示。该模型包括两个部分,在模型的第一个部分,使用 Hotelling T² 控制图对整个过程的质量状态进行判断。当 Hotelling T² 控制图发出过程均值矢量异常报警信号,处于模型第二个部分的 LS-SVM 模式识别器以模式识别的方式辨识和定位过程异常变量。最后,根据 LS-SVM 模式识别器的辨识结果,对引起过程异常的变量(或者其组合)采取纠正措施,使过程恢复到稳定受控状态。

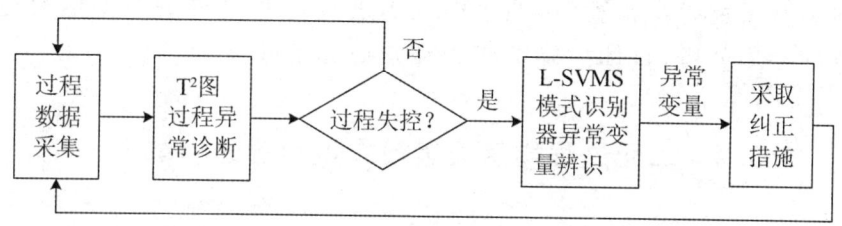

图 5-1 多元过程均值矢量异常诊断与变量辨识智能模型

图 5-1 中的模型将过程异常变量的辨识问题转化为模式识别问题来加以解决。对于一个 P 维制造过程,其每个变量的均值有正常和异常两种状态,因此,P 维过程的均值矢量共有 2^P 种可能的正常/异常状态,其中只有一种正常状态,有 (2^P-1) 种异常状态。一旦 T² 控制图发现过程处于异常状态,则该异常状态必然属于这 (2^P-1) 种异常状态之一。为了将这 2^P 种过程状态加以区分,该诊断模型将过程均值矢量的 (2^P-1) 种不同过程异常状态定义为 (2^P-1) 种需要加以识别的异常模式,并对 LS-SVM 模式识别器进行训练。当 T² 控制图发出异常报警信号后,使用经过训练的 LS-SVM 模式识别器对 (2^P-1) 均值矢量的异常模式进行模式识别,从而实现对过程异常变量的辨识与定位,为下一步采取纠正措施提供必要的信息。以二元过程为例,其均值矢量为 $\mu=(\mu_1,\mu_2)^T$,那么均值矢量只有三种可能的异常状态需要加以辨识,(a)第一个变量异常,第二个变量受控;(b)第一个变量受控,第二个变量异常;(c)两个变量同时发生异常。如果将过程变量的异常和正常状态分别表示为 1 和 0,则二元过程均值矢量的 4 模式可以分别使用 $(0,0)$、$(0,1)$、$(1,0)$ 和 $(1,1)$ 来表示,其中,

(0,0)表示过程均值矢量的正常模式,(0,1)、(1,0)和(1,1)表示过程均值矢量的三种异常模式。当 T^2 控制图发出二元过程异常报警信号,如果 LS-SVM 模式识别器将该异常模式判断为(0,1),表明过程的第一个变量没有发生异常,第二个变量发生了异常;如果 LS-SVM 模式识别器将该异常模式判断为(1,1),表明过程的两个变量都发生了异常。可见 LS-SVM 模式识别器的作用不是取代 Hotelling T^2 控制图,而是当 Hotelling T^2 控制图发出过程均值矢量异常报警时,通过模式识别技术对过程的数据信息进行进一步分析,具体准确地辨识和定位异常变量。

5.3.2 二元过程均值矢量模式识别器的训练

使用图 5-1 中的模式识别器对均值矢量的异常模式进行识别,首先必须使用合适的训练数据对 LS-SVM 识别器进行训练,使其具有模式识别的功能。我们仅就二元过程,说明如何通过仿真的方法生成训练 LS-SVM 识别器所需要数据集。

令 $\mu = (\mu_1, \mu_2)^T$ 和 $\Sigma = \begin{bmatrix} \sigma_1^2 & \rho\sigma_1\sigma_2 \\ \rho\sigma_1\sigma_2 & \sigma_2^2 \end{bmatrix}$ 分别表示具有两个变量 x_1 和 x_2 的二元过程在稳定受控状态下的均值矢量和协方差矩阵,σ_1^2 和 σ_2^2 分别表示两个质量特性变量 x_1 和 x_2 的方差,并且 $x_1 \sim N(\mu_1, \sigma_1^2)$,$x_2 \sim N(\mu_2, \sigma_2^2)$。$\rho$ 为 x_1 和 x_2 之间的相关系数,当过程的均值矢量发生变化时它们的取值保持恒定。使用 M 表示训练 LS-SVM 模式识别器所需的每种模式的样本容量,为 MATLAB 仿真编程矩阵运算方便并不失一般性,M 的数值需取为 4 的整数倍。同时,令 K_{11}、K_{12}、K_{21} 和 K_{22} 表示服从均匀分布的随机数,并且 K_{11} 和 $K_{12} \in [-K_1, K_1]$,K_{31} 和 $K_{32} \in [-K_3, K_3]$,K_{21} 和 $K_{22} \in [-K_2, K_2]$,其中,$K_1 = 0.3$,$K_2 = 1.5$,$K_3 = 3$,则可以通过下面的步骤生成仿真实验中所需的训练数据集。

(1)使用公式 $(\mu_1 + K_{31}\sigma_1, \mu_2 + K_{32}\sigma_2)^T$ 生成 M 个均值矢量作为二元过程均值矢量正常模式的训练数据集,对应的均值矢量正常模式以符号$(0,0)$为标记。此时,应当使用T^2统计量作为约束条件,以保证生成的这M个数据使得过程的均值矢量处于正常模式。

(2)分别使用公式$(\mu_1+K_{11}\sigma_1,\mu_2+K_{21}\sigma_2)^T$和$(\mu_1+K_{12}\sigma_1,\mu_2-K_{22}\sigma_2)^T$各生成$M/2$个均值矢量作为二元过程均值矢量异常模式的训练数据集,对应的均值矢量正常模式以符号$(0,1)$为标记。此时,同样需要使用T^2统计量作为约束条件,以保证生成的这M个数据使得过程的均值矢量处于异常模式。

(3)分别使用公式$(\mu_1+K_{21}\sigma_1,\mu_2+K_{11}\sigma_2)^T$和$(\mu_1-K_{22}\sigma_1,\mu_2+K_{12}\sigma_2)^T$各生成$M/2$个均值矢量作为二元过程均值矢量异常模式的训练数据集,对应的均值矢量正常模式以符号$(1,0)$为标记。同样需要使用T^2统计量作为约束条件,以保证生成的这M个数据使得过程的均值矢量处于异常模式。

(4)分别使用公式$(\mu_1+K_{21}\sigma_1,\mu_2+K_{22}\sigma_2)^T$、$(\mu_1+K_{21}\sigma_1,\mu_2-K_{22}\sigma_2)^T$、$(\mu_1-K_{21}\sigma_1,\mu_2+K_{22}\sigma_2)^T$和$(\mu_1-K_{21}\sigma_1,\mu_2-K_{22}\sigma_2)^T$生成$M/4$个均值矢量作为二元过程均值矢量异常模式的训练数据集,对应的均值矢量正常模式以符号$(1,1)$为标记。同样需要使用T^2统计量作为约束条件,以保证生成的这M个数据使得过程的均值矢量处于异常模式。

在测试LS-SVM模式识别器时,测试数据同样采用上面的方法生成。

5.3.3 模型在二元过程中的应用及性能评估

下面的研究以Montgomery(1991)[190]的一个来自化工过程的二元过程实例来说明如何将图5-1提出的模型应用于过程均值异常诊断及异常变量辨识,同时对该方法的诊断效果进行性能评估。其他文献[1]中使用PCA方法对这个实例进行了研究,该例中的两个质量特性变量为X_1和X_2,由它们组成的基准均值矢量$\bar{X}$和基准协方差矩阵S可以通过最初的15个观测数据(样本容

量 $n=5$)得到,

$$\overline{X} = \begin{bmatrix} 10 & 10 \end{bmatrix}^T, \quad S = \begin{bmatrix} 0.7986 & 0.6793 \\ 0.6793 & 0.7343 \end{bmatrix},$$

取 $\alpha=0.05$,通过式(5.2)和式(5.3)可以得到 T^2 控制图的控制限 $UCL_{T^2} = \dfrac{2 \times 16 \times 14}{15 \times 13} \times 3.81 = 8.753, LCL_{T^2} = 0$。

5.3.3.1 训练数据

为了训练 LS-SVM 均值矢量模式识别器,我们使用得到的 T^2 控制图的控制限和 5.3.2 节介绍的程序,使用仿真的方法生成训练数据。使用仿真软件 MATLAB7,最后得到在不同参数条件下的均值矢量训练数据集。作为例证,表 5-1 给出了仿真生成的四种均值矢量模式的训练数据集(样本容量 $M=24$)以及它们各自的 T^2 值,以判断这些数据是否引起过程失控。

表 5-1 二元过程均值矢量模式训练数据集及其 T^2 值($M=24, K_1=0.3, K_2=1, K_3=3$)

模式(0,0)		模式(0,1)		模式(1,0)		模式(1,1)	
training data	T^2	training data	T^2	training data	T^2	training data	T^2
$(10.96,10.14)^T$	4.05	$(10.11,12.04)^T$	24.27	$(12.56,9.68)^T$	48.08	$(12.63,12.06)^T$	8.85
$(9.13,9.11)^T$	1.09	$(10.23,12.03)^T$	21.43	$(12.45,9.99)^T$	35.4687	$(10.91,12.51)^T$	20.43
$(9.42,8.49)^T$	6.97	$(9.64,12.01)^T$	34.33	$(12.15,9.78)^T$	32.59	$(11.04,12.44)^T$	16.87
$(10.57,11.01)^T$	2.14	$(10.36,11.88)^T$	16.06	$(12.59,10.39)^T$	29.19	$(12.63,12.40)^T$	8.82
$(10.96,11.90)^T$	8.65	$(9.66,12.39)^T$	46.28	$(11.51,9.95)^T$	14.29	$(12.68,11.17)^T$	16.82
$(11.97,11.44)^T$	5.21	$(9.65,12.19)^T$	39.78	$(12.17,9.88)^T$	30.41	$(12.55,11.30)^T$	12.95
$(9.10,9.51)^T$	1.51	$(10.31,12.16)^T$	23.10	$(12.59,9.97)^T$	40.44	$(12.50,8.89)^T$	75.04
$(12.25,11.43)^T$	7.84	$(9.62,12.25)^T$	42.44	$(11.80,9.92)^T$	20.59	$(11.41,8.61)^T$	45.32
$(10.95,10.58)^T$	1.43	$(10.32,12.23)^T$	24.57	$(11.98,10.37)^T$	15.92	$(12.02,8.28)^T$	80.48
$(10.11,9.83)^T$	0.47	$(9.59,11.83)^T$	30.59	$(11.75,10.24)^T$	13.71	$(11.15,7.98)^T$	58.98

续表

模式(0,0)		模式(0,1)		模式(1,0)		模式(1,1)	
training data	T^2	training data	T^2	training data	T^2	training data	T^2
$(12.03,11.26)^T$	6.50	$(10.12,12.52)^T$	37.52	$(11.30,9.85)^T$	12.16	$(11.52,8.91)^T$	39.25
$(9.35,8.79)^T$	3.30	$(9.59,11.90)^T$	32.60	$(11.85,9.51)^T$	31.73	$(12.39,9.08)^T$	62.70
$(11.66,11.11)^T$	4.04	$(10.49,7.90)^T$	40.68	$(8.50,9.89)^T$	11.58	$(8.18,11.00)^T$	83.91
$(9.47,10.11)^T$	2.34	$(10.03,7.49)^T$	41.12	$(7.33,9.94)^T$	40.14	$(7.90,12.21)^T$	107.54
$(11.77,11.02)^T$	5.45	$(9.58,7.87)^T$	20.21	$(8.18,9.62)^T$	12.87	$(7.82,12.03)^T$	102.33
$(8.44,9.34)^T$	5.83	$(10.36,8.97)^T$	11.53	$(7.41,9.84)^T$	35.09	$(8.03,11.79)^T$	81.25
$(8.56,8.79)^T$	2.59	$(10.47,7.49)^T$	54.10	$(9.04,10.32)^T$	9.45	$(8.72,11.28)^T$	37.89
$(12.06,12.11)^T$	6.11	$(10.02,8.70)^T$	11.04	$(7.85,10.42)^T$	38.05	$(8.92,12.29)^T$	67.08
$(11.66,11.80)^T$	4.40	$(10.09,8.68)^T$	12.43	$(8.56,10.01)^T$	12.41	$(8.29,7.61)^T$	9.27
$(8.52,8.05)^T$	5.80	$(9.79,8.63)^T$	9.20	$(7.74,9.64)^T$	22.14	$(7.48,9.12)^T$	18.14
$(7.84,8.26)^T$	5.88	$(10.50,8.77)^T$	17.79	$(7.37,9.69)^T$	32.34	$(9.02,7.75)^T$	14.04
$(8.83,9.98)^T$	7.67	$(10.19,7.76)^T$	36.90	$(7.62,9.60)^T$	24.03	$(8.93,7.82)^T$	11.79
$(11.27,10.75)^T$	2.67	$(10.47,8.35)^T$	27.16	$(7.34,9.73)^T$	34.19	$(7.58,9.06)^T$	15.40
$(9.51,9.13)^T$	1.59	$(9.82,8.59)^T$	10.10	$(7.34,9.67)^T$	32.76	$(7.34,8.07)^T$	9.56

根据表 5-1 容易验证,所有正常模式的训练数据的 T^2 值都没有超过 UCL_{T^2}(8.753),而所有异常模式的训练数据的 T^2 值都超过了 UCL_{T^2}(8.753)。

为了比较不同大小的训练样本容量 M 对于模型诊断性能的影响,实验在不同的 M 的取值下进行,M 以递增的方式取 $M=8$,16,24,36,44,52。根据概率统计学的一般原理,这里认为 $M=8$,16,24 时,样本容量较小;$M=36,44,52$ 时,样本容量较大。而且,实验也分别研究了诊断模型在过程中存在较小的过程随机扰动(仿真取 $K_1=0.3$)和较大的过程随机扰动(仿真取 $K_1=0.6$)情况下的性能表现。图 5-2 用以说明训练数据的分布情况,对应于二元过程均值矢量的四种模式(0,0)、(0,1)、(1,0)和(1,1),在图 5-2 中分别使用符号(○)、(∗)、(×)和(▽)进行标记。

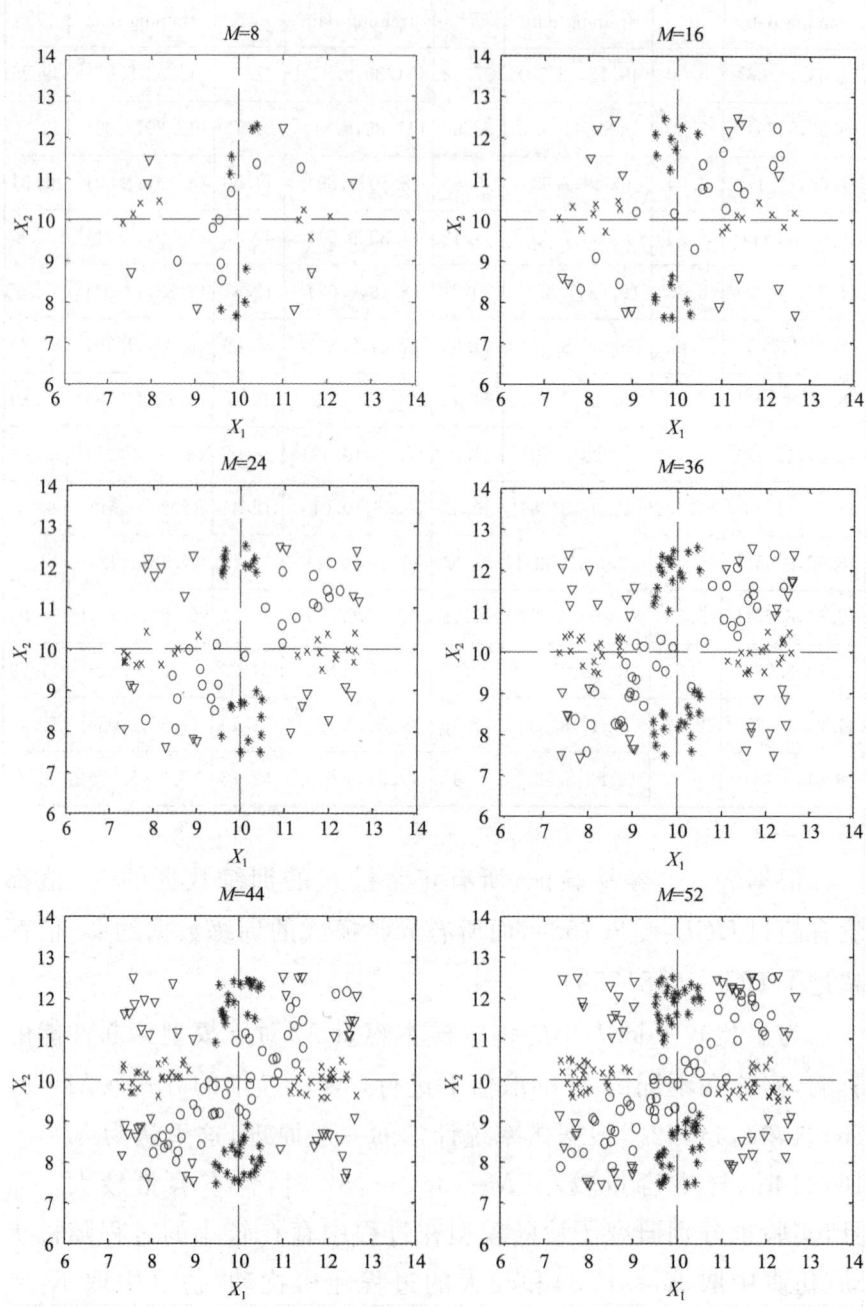

图 5-2　训练数据分布图($K_1=0.3, K_2=1, K_3=3$)

根据图 5-2 的数据分布图和 SVM 理论原理,不同模式的数据在高维特征空间可以被最优分类超平面正确分类。另外,根据多元统计学,当使用 T^2 统计量作为约束条件来生成均值矢量的 4 类模式时,属于正常模式的均值矢量 (X_1,X_2) 应当位于 X_1-X_2 平面的某个椭圆区域内,属于异常模式的均值矢量 (X_1,X_2) 应当位于 X_1-X_2 平面的某个椭圆区域外,这一特点在图 5-2 中,当样本容量较大时表现得更为明显。

5.3.3.2 仿真实验及性能评估

生成训练数据集之后,下面的实验首先要使用训练数据集和"一对一"算法训练 LS-SVM 模式识别器,并通过调整优化 LS-SVM 模式识别器的参数 σ 和 γ,使其对 4 种均值矢量模式具有良好的识别率性能。对于均值矢量的四种模式 $(0,0)$、$(0,1)$、$(1,0)$ 和 $(1,1)$,在 LS-SVM 理论中使用符号 $y_1=1, y_2=2, y_3=3, y_4=4$ 来表示。测试 LS-SVM 模式识别器性能时,每类模式各使用 1000 个测试数据。实验过程中,通过"交叉验证法"得到优化的 LS-SVM 模式识别器参数 $\gamma=0.15, \sigma=0.3$。仿真实验使用的数据参数及实验结果总结在表 5-2 和表 5-3 中。

表 5-2 过程随机扰动较小时 LS-SVM 模式识别器的性能

训练数据参数		测试数据参数	模式类型				识别性能(%)		
M	(K_1,K_2,K_3)	$(K_1=0.3, K_2=1, K_3=3)$	$(0,0)$	$(0,1)$	$(1,0)$	$(1,1)$	识别率	误判率	漏判率
8	$(0.3,1,3)$	$(0,1)$	16	984	0	0	98.4	0	1.6
		$(1,0)$	5	0	993	2	99.3	0.2	0.5
		$(1,1)$	168	7	39	786	78.6	4.6	16.8
	$(0.6,1,3)$	$(0,1)$	33	967	0	0	96.7	0	3.3
		$(1,0)$	0	0	1000	0	100	0	0
		$(1,1)$	78	105	5	812	81.2	11	7.8

续表

训练数据参数 M	训练数据参数 (K_1, K_2, K_3)	测试数据参数 $(K_1=0.3, K_2=1, K_3=3)$	模式类型 (0,0)	模式类型 (0,1)	模式类型 (1,0)	模式类型 (1,1)	识别性能(%) 识别率	识别性能(%) 误判率	识别性能(%) 漏判率
16	(0.3,1,3)	(0,1)	0	1000	0	0	100	0	0
16	(0.3,1,3)	(1,0)	0	0	1000	0	100	0	0
16	(0.3,1,3)	(1,1)	14	3	39	944	94.4	4.2	1.4
16	(0.6,1,3)	(0,1)	0	1000	0	0	100	0	0
16	(0.6,1,3)	(1,0)	0	0	998	2	99.8	0.2	0
16	(0.6,1,3)	(1,1)	46	32	1	921	92.1	3.3	4.6
24	(0.3,1,3)	(0,1)	0	1000	0	0	100	0	0
24	(0.3,1,3)	(1,0)	0	0	1000	0	100	0	0
24	(0.3,1,3)	(1,1)	43	6	9	942	94.2	1.5	4.3
24	(0.6,1,3)	(0,1)	0	1000	0	0	100	0	0
24	(0.6,1,3)	(1,0)	18	0	982	0	98.2	0	1.8
24	(0.6,1,3)	(1,1)	15	18	23	944	94.4	4.1	1.5
36	(0.3,1,3)	(0,1)	0	1000	0	0	100	0	0
36	(0.3,1,3)	(1,0)	0	0	1000	0	100	0	0
36	(0.3,1,3)	(1,1)	25	0	0	975	97.5	0	2.5
36	(0.6,1,3)	(0,1)	0	1000	0	0	100	0	0
36	(0.6,1,3)	(1,0)	0	0	1000	0	100	0	0
36	(0.6,1,3)	(1,1)	12	7	17	964	96.4	2.4	1.2
44	(0.3,1,3)	(0,1)	0	1000	0	0	100	0	0
44	(0.3,1,3)	(1,0)	0	0	1000	0	100	0	0
44	(0.3,1,3)	(1,1)	9	1	1	989	98.9	0.2	0.9
44	(0.6,1,3)	(0,1)	0	1000	0	0	100	0	0
44	(0.6,1,3)	(1,0)	0	0	1000	0	100	0	0
44	(0.6,1,3)	(1,1)	1	1	19	979	97.9	2	0.1

续表

训练数据参数		测试数据参数	模式类型				识别性能(%)		
M	(K_1,K_2,K_3)	$(K_1=0.3, K_2=1,K_3=3)$	(0,0)	(0,1)	(1,0)	(1,1)	识别率	误判率	漏判率
52	(0.3,1,3)	(0,1)	0	1000	0	0	100	0	0
		(1,0)	0	0	1000	0	100	0	0
		(1,1)	10	0	2	988	98.8	0.2	1
	(0.6,1,3)	(0,1)	0	1000	0	0	100	0	0
		(1,0)	0	0	1000	0	100	0	0
		(1,1)	0	4	19	977	97.7	2.3	0

表 5-3 过程随机扰动较大时 LS-SVM 模式识别器的性能

训练数据参数		测试数据参数	模式类型				识别性能(%)		
M	(K_1,K_2,K_3)	$(K_1=0.6, K_2=1,K_3=3)$	(0,0)	(0,1)	(1,0)	(1,1)	识别率	误判率	漏判率
8	(0.3,1,3)	(0,1)	45	906	0	49	90.6	4.9	4.5
		(1,0)	0	0	1000	0	100	0	0
		(1,1)	101	36	143	720	72	17.9	10.1
	(0.6,1,3)	(0,1)	0	1000	0	0	100	0	0
		(1,0)	13	0	975	12	97.5	1.2	1.3
		(1,1)	27	60	20	893	89.3	8	2.7
16	(0.3,1,3)	(0,1)	2	998	0	0	99.8	0	0.2
		(1,0)	42	0	958	0	95.8	0	4.2
		(1,1)	35	38	19	908	90.8	5.7	3.5
	(0.6,1,3)	(0,1)	0	1000	0	0	100	0	0
		(1,0)	0	0	984	16	98.4	1.6	0
		(1,1)	12	56	28	904	90.4	8.4	1.2

续表

训练数据参数		测试数据参数 $(K_1=0.6,$ $K_2=1,K_3=3)$	模式类型				识别性能(%)		
M	(K_1,K_2,K_3)		(0,0)	(0,1)	(1,0)	(1,1)	识别率	误判率	漏判率
24	(0.3,1,3)	(0,1)	14	986	0	0	98.6	0	1.4
		(1,0)	1	0	999	0	99.9	0	0.1
		(1,1)	18	12	23	947	94.7	3.5	1.8
	(0.6,1,3)	(0,1)	0	997	0	3	99.7	0.3	0
		(1,0)	0	0	1000	0	100	0	0
		(1,1)	43	6	3	948	94.8	0.9	4.3
36	(0.3,1,3)	(0,1)	7	977	0	16	97.7	1.6	0.7
		(1,0)	0	0	997	3	99.7	0.3	0
		(1,1)	26	2	5	967	96.7	0.7	2.6
	(0.6,1,3)	(0,1)	2	998	0	0	99.8	0	0.2
		(1,0)	0	0	1000	0	100	0	0
		(1,1)	15	9	12	964	96.4	2.1	1.5
44	(0.3,1,3)	(0,1)	0	1000	0	0	100	0	0
		(1,0)	0	0	1000	0	100	0	0
		(1,1)	16	0	0	984	98.4	0	1.6
	(0.6,1,3)	(0,1)	0	1000	0	0	100	0	0
		(1,0)	0	0	1000	0	100	0	0
		(1,1)	20	0	15	965	96.5	1.5	2
52	(0.3,1,3)	(0,1)	0	1000	0	0	100	0	0
		(1,0)	0	0	1000	0	100	0	0
		(1,1)	0	0	10	990	99	1	0
	(0.6,1,3)	(0,1)	0	1000	0	0	100	0	0
		(1,0)	0	0	1000	0	100	0	0
		(1,1)	0	0	2	998	99.8	0.2	0

表 5-2 和表 5-3 两个表的第 1 列为每种模式训练数据的样本

容量,第2列为产生这些训练数据所使用的参数;第3列表明了生成的1000个测试数据的类别,这两个表的第3列的第2行表示生成这些测试数据使用的参数;两个表的第4~7列表示了某个类别的1000个测试数据被分类的情况;两个表的第8~10列给出了评价该诊断模型性能的技术指标(即,识别率、误判率和漏判率)。表5-2和表5-3的参数K_1用以表示和控制生成的训练/测试数据中普通原因变差的大小。两个表的区别在于它们第3列的参数K_1的取值不同,该值在表5-2中$K_1=0.3$,表示在测试数据中包含较小的普通原因变差;该值在表5-3中$K_1=0.6$,表示在测试数据中包含较大的普通原因变差。

根据表5-2和表5-3中的实验结果并通过对它们进行比较,可以得出下面的实验结论:

(1)训练数据的样本容量对诊断性能的影响。增大训练数据的样本容量可以提高模型对于均值异常模式的识别率。由于每种模式的训练样本容量为16或24时,模型对于异常模式的识别率已经高于90%,因此我们认为提出的诊断模型在训练样本容量较小时就具有了很好的异常诊断性能。另一方面,当每种模式的训练样本容量大于36时,进一步提高训练样本容量并不能明显提高模型的诊断性能,反而会增加质量成本。

(2)过程中的随机扰动对于模型的诊断性能的影响。首先,当训练数据中的扰动增大时(即,生成训练数据时使用的K_1值由0.3增大到0.6,而其他参数不变),模型的识别率性能没有发生大的变化。其次,当测试数据中的扰动增大时(即,生成测试数据时使用的K_1值由0.3增大到0.6,而其他参数不变),诊断模型对于异常模式的识别率仅仅发生很微小的下降,这种变化完全可以忽略。因此诊断模型具有较好的抗随机干扰能力。

(3)LS-SVM模式识别器的参数γ和σ对模型的诊断性能有很大的影响,需要优化选择。

另外可以看到,诊断模型对于(1,1)型均值矢量异常模式的识别率低于诊断模型对于(0,1)型和(1,0)型均值矢量异常模式的识别率。

5.3.3.3 诊断实例比较

Jackson(1991)[191]对于同样研究了上面的二元过程工业案例,并且指出过程的 4 个均值矢量观测值 A=(12.3,12.5)^T、B=(7.0,7.3)^T、C=(11.0,9.0)^T 和 D=(7.3,9.1)^T 全部处于异常状态,但无法说明究竟是哪个变量引起过程均值异常。我们使用在参数 $M = 52$、$K_1 = 0.6$、$K_2 = 1$、$K_3 = 3$ 下训练得到的 LS-SVM 模式识别器对上述四个均值矢量进行诊断,四个均值矢量观测数据被判定都属于(1,1)模式,这一诊断结论与 Jackson 的诊断结果相符合,而且诊断结果说明了这两个变量都是引起过程异常的原因。

通过上面智能过程异常诊断模型在二元过程中的一个应用实例,虽然不能代表智能诊断模型在各种制造过程条件下对于过程均值异常的诊断能力,但仍表现出了智能诊断模型的应用特点和技术优势。

5.4 多元过程散度异常诊断与变量识别的智能模型

对于多元制造过程,除了需要监控过程均值矢量的异常状况,过程的散度异常变化同样需要加以监控。对于一个 P 维多元过程,过程的散度一般使用 $P \times P$ 的协方差矩阵 Σ 来描述。当监控和诊断多元过程的散度变化时,单纯使用传统 MSPC 控制图只能对过程的总体状况进行判断,不能具体说明究竟是哪个过程变量(或者哪几个过程变量的组合)造成了过程异常,仍然存在着需要辨识过程异常变量的问题。多元过程协方差矩阵 Σ 的对角线上的元素为各个质量特性变量的方差,其非对角线上的元素为质量特性变量之间的协方差,由于方差和协方差都是由过程原始观测数据经过计算得到的,过程的故障特征存在被掩盖的可能,因此,相对于诊断和监控多元过程均值的异常变化,诊断多元过程散度异常变化并进行异常变量辨识的工作难度将会增加。

5.4.1 多元过程协方差矩阵异常的智能诊断模型

MSPC 控制图与单变量 SPC 控制图一样,同样分为在过程的第一个阶段使用的分析用 MSPC 控制图和在过程的第二个阶段使用的控制用 MSPC 控制图[192]。在多元过程的分析阶段,质量工程师需要分析多元过程是否受控,并在过程达到受控状态后建立过程的统计受控标准,并确定多元过程均值和协方差矩阵的基准值,即 (μ_0,Σ_0);在多元过程的控制阶段,MSPC 控制图用来监控过程中是否出现异常,使得多元过程偏离了其基准参数 (μ_0,Σ_0)。针对诊断多元过程协方差矩阵异常变化问题,我们首先假定过程中的异常因素仅仅使过程的协方差矩阵发生异常变化,不会影响到过程的均值矢量和变量间的相关系数,然后,我们在制造过程的第二个阶段构建过程异常诊断的智能模型,在使用过程的总体统计量监控过程协方差矩阵异常变化的同时,使用 LS-SVM 模式识别技术实现对过程异常变量的辨识与定位,模型的结构见图 5-3。

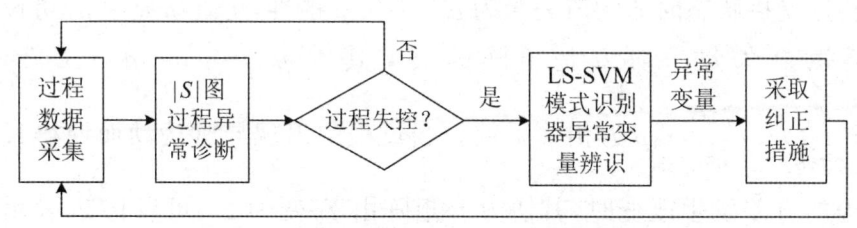

图 5-3 多元过程协方差矩阵异常诊断与变量辨识智能模型

图 5-3 中的智能模型由两部分组成:方差异常检测器(detector)和 LS-SVM 模式识别器。在模型的第一部分,使用样本广义方差 $|S|$ 图作为方差异常检测器连续地监控过程,以诊断过程的

方差是否出现异常；当 |S| 图发出过程异常报警信号，LS-SVM 模式识别器将对过程的协方差矩阵进行分析，以模式识别的方式来辨识和确定究竟是哪个过程变量引起了过程方差异常。样本广义方差 |S| 控制图技术实际上是统计学假设检验的一种图形化表达方法，使用 |S| 图诊断过程协方差矩阵的异常变化时，需要假定当多元过程中出现异常扰动时，过程的均值矢量和质量特性变量之间的相关系数仍然保持不变。另一方面，用来辨识过程异常变量的 LS-SVM 模式识别技术不同于多元统计学中的过程异常诊断技术与方法，多元过程诊断模型中 LS-SVM 等智能技术的使用，其目的不是要用来取代传统的 MSPC 技术，而是要用来补充和增强后者的功能。

5.4.2 二元过程协方差矩阵模式的定义

二元过程虽然是最简单的多元过程，但是对于二元过程的研究却是研究二元以上多元过程的基础，很多对于多元过程的研究思路都是对二元过程研究思路的拓展和改进。因此我们仍然仅就二元过程来说明如何使用图 5-3 中的模型来诊断过程协方差矩阵出现的异常并且实现过程异常变量的辨识。

对于一个具有两个变量的二元过程 $X=(x_1,x_2)$，假定其处于稳定受控状态时的均值矢量为 μ_0（不失一般性，仿真研究中 μ_0 可设定为 0 矢量）、协方差矩阵为 Σ_0，其中，$\mu_0 = (\mu_{10}, \mu_{20})$，$\Sigma_0 = \begin{bmatrix} \sigma_1^2 & \rho\sigma_1\sigma_2 \\ \rho\sigma_1\sigma_2 & \sigma_2^2 \end{bmatrix}$，$\rho$ 为相关系数。当过程中出现异常扰动而使得过程的方差发生改变时，其协方差矩阵由 Σ_0 变为 Σ_1，可以将 Σ_1 表示为：

$$\Sigma_1 = \begin{bmatrix} a^2\sigma_1^2 & \rho ab\sigma_1\sigma_2 \\ \rho ab\sigma_1\sigma_2 & b^2\sigma_2^2 \end{bmatrix} \quad (5.13)$$

令 $d^2 = |\Sigma_1|/|\Sigma_0|$，则 d^2 表示失控过程的协方差矩阵与受控过程的协方差矩阵的"行列式比"，引起过程失控的扰动信号的大

小强弱可以使用 d^2 来表示。根据 Σ_1 的表达式(5.13)，协方差矩阵 Σ_1 的正常模式和异常模式可以使用下面的方法进行定义。

(1)第一种模式为正常模式，过程中没有发生异常，此时 $a = b = d = 1$，将正常模式用符号(0,0)来表示。

(2)第二种形式的异常模式表示过程变量 x_1 的方差受到异常因素的影响而发生异常改变，过程变量 x_2 的方差没有受到过程中的异常因素影响而发生异常改变，此时，$a^2 = d^2 > 1$ 并且 $b = 1$，将这种类型的异常模式用符号(1,0)来表示。

(3)第三种形式的异常模式表示过程变量 x_2 的方差受到异常因素的影响而发生异常改变，过程变量 x_1 的方差没有受到过程中的异常因素影响而发生异常改变，此时，$b^2 = d^2 > 1$ 并且 $a = 1$，将这种类型的异常模式用符号(0,1)来表示。

(4)第四种形式的异常模式表示过程变量 x_2 和 x_1 的方差都受到异常因素的影响而发生异常改变，此时，$a^2 = b^2 = d^2 > 1$，将这种类型的异常模式用符号(1,1)来表示。

通过对二元过程的协方差矩阵 Σ_1 进行模式定义，就可以将统计学协方差矩阵的异常诊断问题转化为对协方差矩阵的模式识别问题，从而使用 LS-SVM 模式识别技术来诊断和辨识过程中的异常变量。在训练 LS-SVM 模式识别器时，将协方差矩阵模式的数目设置为4，即 $y_i \in \{1,2,3,4\}$。

5.4.3 二元过程方差异常智能诊断模型的仿真研究

为了使 LS-SVM 模式识别器具有识别二元过程协方差矩阵模式的功能，必须使用相应的数据对 LS-SVM 模式识别器进行训练。获得训练数据的最好方法是结合具体的质量诊断实际问题从过程现场采集数据，但由于成本、时间等诸多条件的限制，这种数据的获取方法难以实现。此时，可以使用仿真的方法作为一种替代方法来研究制造过程的异常诊断问题，也同样可以取得良好的效果[193]。因此我们仍然使用仿真的方法来生成训练 LS-SVM 模式识别器所需的各类协方差矩阵模式数据。

5.4.3.1 原始过程仿真数据的生成

根据前面的内容,对于二元过程使用公式(5.13),通过控制变量 a 和 b 的取值,就可以为每种类型的协方差矩阵模式生成 N 组过程仿真观测数据 $X_{1k}, X_{2k}, \cdots, X_{nk}, k = 1, 2, \cdots, N$。其中,$n$ 表示在第 k 次采样时的样本子组容量。本书的研究中 a、b 的值要么在区间 $[1.5, 3]$ 内随机生成,要么设置为 1。比如,为了生成 $(0, 1)$ 类型的协方差矩阵模式,我们将 a 的值设为 1,b 的值在区间 $[1.5, 3]$ 内随机生成。为了得到满足均值矢量为 μ_0、协方差矩阵为 Σ_0 的正态分布多元过程仿真数据,在使用 MATLAB 软件进行仿真时,用到函数"mvnrnd"。同时,在生成训练数据的过程中,必须使用 $|S|$ 图作为约束条件,以保证生成的所有用于协方差矩阵正常模式的训练数据的广义方差($|S|$ 值)没有超过 $|S|$ 图的控制限,而所有用于协方差矩阵异常模式的训练数据的广义方差($|S|$ 值)超过了 $|S|$ 图的控制限。

5.4.3.2 LS-SVM 模式识别器的输入特征矢量

在为每种类型的协方差矩阵模式生成 N 组过程仿真观测数据 $X_{1k}, X_{2k}, \cdots, X_{nk}, k = 1, 2, \cdots, N$,这些数据并不能直接用于 LS-SVM 模式识别器的训练,因为如果不从这些数据中提取合适的特征矢量来训练 LS-SVM 模式识别器,就不能使 LS-SVM 模式识别器获得良好的模式识别性能。

设 $X_{1k}, X_{2k}, \cdots, X_{nk}$ 表示在第 k 次采样时的过程样本数据,n 为第 k 次采样时的样本子组容量,我们在研究中使用下面的公式来构造 LS-SVM 模式识别器的输入特征矢量 V_k:

$$V_k = (S_{1k}^2, S_{12k}, S_{2k}^2) \tag{5.14}$$

其中,S_{12k} 表示两个质量特征变量 x_1 和 x_2 的样本协方差,S_{1k}^2 和 S_{2k}^2 分别为两个质量特征变量 x_1 和 x_2 各自的样本方差。对于某种协方差矩阵模式,需要经过 N 次采样,共获得 N 个输入特征矢量 $V_k, k=1,2,\cdots,N$,用于该类协方差矩阵模式的训练。由于我们需要监控二元过程散度的异常变化,故选用过程变量的方差和协方差作为 LS-SVM 识别器的输入特征矢量元素。

5.4.3.3 诊断模型仿真研究与性能评估

我们使用二元过程的例子通过仿真实验来评价图 5-3 中提出的诊断模型的性能及有效性。为了简单起见并不失一般性,研究中假定二元过程的两个变量 x_1 和 x_2 服从均值矢量为零、方差为 1、相关系数为 ρ 的二维联合正态分布。在实际的应用中,质量特性变量的值可以通过公式 $\dfrac{X-\mu}{\sigma}$ 进行变换以满足上述假设。

在实验过程中,对于协方差矩阵的四种模式,其训练数据和测试数据的生成方法是相同的,并且在同一次实验中,训练数据和测试数据的样本子组容量也是相同的。LS-SVM 模式识别器的参数 γ 和 σ 需要优化选择,最后得到 $\gamma=1, \sigma=0.3$。仿真实验中,使用"一对一"算法建立多类模式分类器,每类模式使用的训练样本容量为 $N=15$,样本子组容量为 $n=10$,以及不同的相关系数 ρ,最后得到的实验结果如表 5-4 所示。

表 5-4 LS-SVM 模式识别器性能

(训练样本容量为 $N=15$,样本子组容量为 $n=10$)

| ρ | 1000 个测试矢量数据的模式类别 | \multicolumn{4}{c}{1000 个测试矢量数据的分类情况} | \multicolumn{3}{c}{模式识别性能(%)} |
		(0,0)	(0,1)	(1,0)	(1,1)	识别率	误判率	漏判率
0.2	(0,1)	9	866	12	113	86.6	12.5	0.9
	(1,0)	7	1	948	44	94.8	4.5	0.7
	(1,1)	2	140	167	691	69.1	30.7	0.2

续表

ρ	1000个测试矢量数据的模式类别	1000个测试矢量数据的分类情况				模式识别性能（%）		
		(0,0)	(0,1)	(1,0)	(1,1)	识别率	误判率	漏判率
0.4	(0,1)	38	911	35	16	91.1	5.1	3.8
	(1,0)	38	0	783	179	78.3	17.9	3.8
	(1,1)	11	141	166	682	68.2	30.7	1.1
0.6	(0,1)	18	884	64	34	88.4	9.8	1.8
	(1,0)	4	2	888	106	88.8	10.8	0.4
	(1,1)	2	200	178	620	62.0	37.8	0.2
0.8	(0,1)	15	903	0	82	90.3	08.2	1.5
	(1,0)	24	7	854	115	85.4	12.2	2.4
	(1,1)	3	135	96	766	76.6	23.1	0.3

从表 5-4 可知,当变量 x_1 和 x_2 之间的相关系数在很大的范围变化时,LS-SVM 模式识别器的性能不受相关系数 ρ 变化的影响。应当注意,上面结论成立的前提是假定过程中的异常扰动并不会改变变量 x_1 和 x_2 之间的相关系数 ρ 以及过程均值,当这一假设前提不成立时,LS-SVM 模式识别器的性能可能会下降。这一问题有待进一步研究。根据表 5-4 还可以看出,在各种实验条件下,LS-SVM 模式识别器对于(0,1)和(1,0)型协方差矩阵模式的识别性能优于 LS-SVM 模式识别器对于(1,1)型协方差矩阵模式的识别性能;LS-SVM 模式识别器对于(0,1)和(1,0)型协方差矩阵模式的识别性能基本上是一样的。

为了进行比较研究,仿真实验在不同的训练样本数 N 和样本子组数 n 下进行,实验结果就如图 5-4 和图 5-5 所示。这两个图表明,选择较大的训练样本数 N 和样本子组数 n 有利于提高 LS-SVM 模式识别器对于各类异常协方差矩阵模式的识别率。

第 5 章　多元过程质量智能诊断与异常变量识别

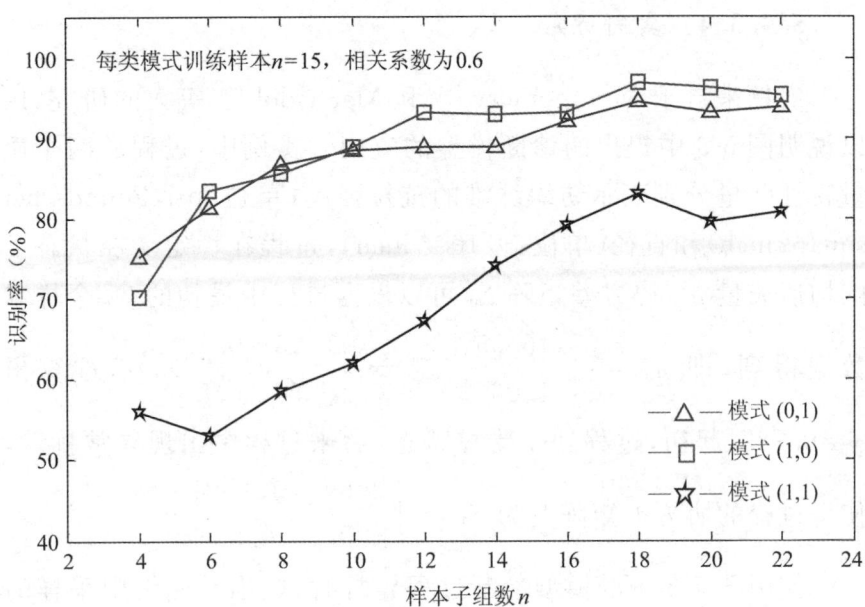

图 5-4　识别率与样本子组数 n 之间的关系图

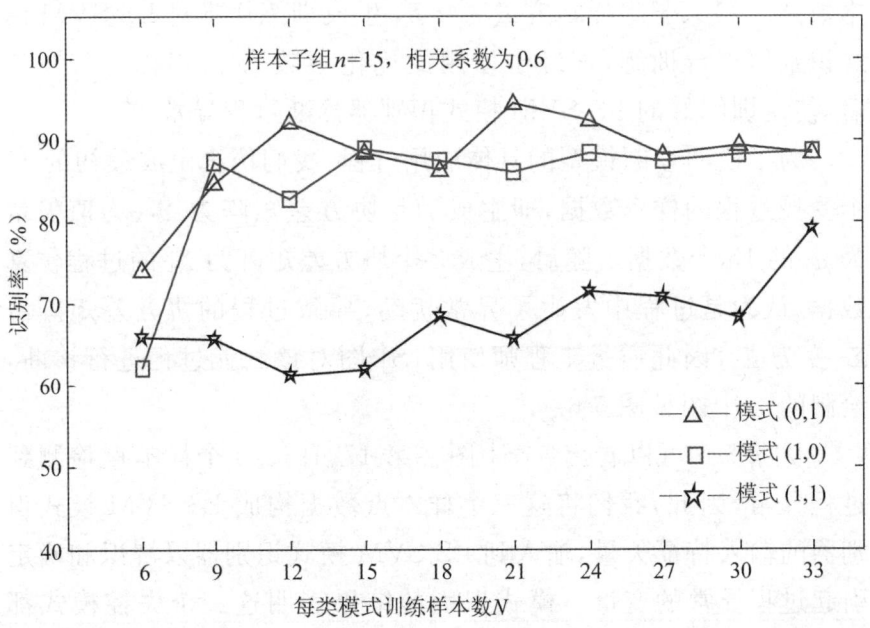

图 5-5　识别率与训练样本数 N 之间的关系图

5.4.3.4 实例研究

实例来自于 Montgomery[194] 和 Machadol[176] 等人的研究,用以说明图 5-3 中提出的诊断模型的应用。实例中,过程的两个质量特性变量分别表示纺织纤维的抗拉强度(单位:psi,Pounds per square inch)和直径(单位:×10^{-2} inch),过程在稳定受控状态下的均值矢量 μ_0、协方差矩阵 Σ_0 可以根据过程中最初的 20 个样本数据得到,即 $\mu_0 = \begin{bmatrix} 115.59 \\ 1.06 \end{bmatrix}$,$\Sigma_0 = \begin{bmatrix} 1.23 & 0.79 \\ 0.79 & 0.83 \end{bmatrix}$,不难算出 $\rho=0.782$。起初,过程处于受控状态,后来过程中出现异常扰动,使得过程的协方差矩阵变为 $\Sigma_1 = \begin{bmatrix} 3.69 & 1.368 \\ 1.368 & 0.83 \end{bmatrix}$。

使用图 5-3 中的模型诊断过程异常时,应当首先设定采样的样本子组容量 $n=6$,则根据 μ_0 和 Σ_0 等已知的过程参数,可以算出 $|S|$ 图的 $UCL=1.316$,$LCL=0$;然后,选择每类模式训练样本数 $N=15$、子组样本容量 $n=6$,生成训练样本对 LS-SVM 模式识别器进行训练,训练中使用的优化参数为 $\gamma=1$,$\sigma=0.3$;最后,使用训练好的 LS-SVM 模式识别器诊断过程异常。

为了说明诊断模型的具体应用过程,我们仿真生成最初的 15 个受控过程的样本数据,即生成满足协方差矩阵为 Σ_0、均值矢量为 μ_0 的 15 个数据。随后,生成多个协方差矩阵为 Σ_1 的过程仿真数据,认为是过程中发生了异常扰动,导致过程的协方差矩阵由 Σ_0 变为 Σ_1,因此质量工程师使用 $|S|$ 图对该二元过程进行诊断,绘制的 $|S|$ 图见图 5-6。

从图 5-6 可以看出,$|S|$ 图在第 17、19、20 个样本点检测到过程异常,因此,我们将这三个样本点数据构成 LS-SVM 模式识别器的输入特征矢量,输入到 LS-SVM 模式识别器以辨识和确定引起过程异常的变量。模式识别的结果表明这三个失控模式都属于(1,0)型失控模式。LS-SVM 模式识别器的诊断结果认为第一个质量特性变量应当为过程的异常和失控负责。也就是说,第

一个质量特性变量的观测值受到了异常扰动的影响而使得其方差发生了异常变化,导致过程失控。因此,基于 LS-SVM 模式识别器的诊断结果,质量工程师需要针对第一个质量特性变量采取纠正措施,使制造过程恢复到稳定受控状态。上面使用 LS-SVM 模式识别器对过程的诊断结果与 Machado 使用统计学方法的诊断结果一致,都认为第一个质量变量是导致过程发生异常的原因。与质量诊断的统计学方法相比,使用 SVM 模式识别技术进行过程诊断,实际上并不需要知道关于过程分布的先验知识,在实际应用中具有更大的适用性。

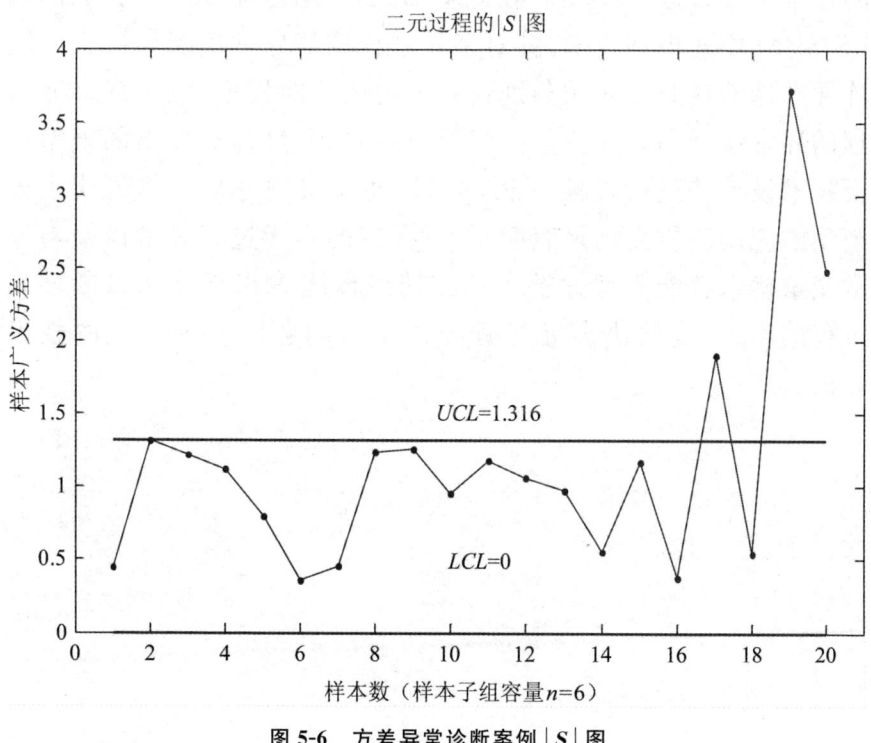

图 5-6　方差异常诊断案例 $|S|$ 图

5.5　本章小结

现代生产条件下产品的制造过程是一个复杂的制造系统,其特点之一就是具有多个质量特性变量并且存在相关性，多变量制造过程的质量诊断问题不能简单地使用单变量 SPC 技术来处理和解决。对多变量制造过程进行异常诊断，不仅要对过程的总体质量状态作出判断，还需要辨识和分离异常的质量特性变量，以便于查找过程异常的原因并实施过程改进措施。

  对多元过程的诊断一般包括过程均值异常诊断和过程协方差矩阵异常诊断两个方面的基本问题。本章首先介绍了典型的多元 SPC 控制图，包括用于过程均值异常诊断的 Hotelling T^2 控制图、MEWMA 图、MCUSUM 图以及用于过程协方差异常诊断的样本广义方差 $|S|$ 图。在此基础上，使用传统 MSPC 控制图和 LS-SVM 模式识别技术，针对多元过程均值异常诊断和协方差矩阵异常诊断这两个问题分别建立了智能诊断模型，以实现多元过程的异常诊断和异常变量的辨识；结合二元过程对提出的模型进行技术设计并进行性能评估，同时，通过实例分析具体说明了该模型的应用。研究结果表明，本章提出的多元过程异常诊断与异常变量辨识智能模型弥补了单纯使用传统 MSPC 技术诊断多元过程的不足，是解决多元过程异常诊断问题的一种有效的技术方法。

第 6 章 总结与展望

6.1 研究内容总结

本书在传统质量诊断技术的基础上,重点研究如何将人工智能技术应用于现代制造过程的质量诊断问题,提出了基于人工智能技术、在不同的制造环境与客户需求条件下进行过程质量诊断的一系列的思路、模型及技术实现方法,主要在以下几个方面进行了研究并取得部分阶段性研究成果。

1)研究了基于概率神经网络(PNN)的控制图模式识别问题

单变量质量控制图是诊断过程异常的重要工具,但单变量SPC控制图的过程异常判断准则并不能包括所有的过程异常现象。本书首先研究了控制图正常/异常模式现象及控制图模式的Monte-Carlo仿真实现方法、仿真数据处理方法;研究了神经网络在控制图模式识别中的应用问题;针对以往使用BP等神经网络识别控制图模式时存在的问题,提出并重点研究使用概率神经网络进行控制图模式识别的技术方法,对概率神经网络的结构进行了研究和设计,对其参数进行了优化设计;提出了计算ARL的移动窗口法,并且基于识别率和ARL指标,对使用概率神经网络识别控制图模式的性能进行了评估,取得良好效果。概率神经网络的训练需要使用大量的样本数据,该方法适用于解决批量生产条件下的过程质量诊断问题。

2)研究了基于最小二乘支持向量机(LS-SVM)的控制图模式识别技术及参数优化问题

多品种小批量生产是先进制造环境下生产过程的显著特征,

制造过程质量异常诊断的智能方法研究

在小批量生产条件下进行过程质量诊断是国民经济和市场发展的客观要求,基于大样本数据的传统统计学质量诊断方法在小样本条件下并不适用。在研究统计学习理论的基础上,本书提出使用最小二乘支持向量机技术实现在有限训练样本条件下进行控制图的模式识别,构建了 LS-SVM 多类模式分类器,通过仿真实验对提出的方法进行了性能评估。针对 LS-SVM 模式分类器的参数优化选择问题,提出使用进化优化算法来优化选择 LS-SVM 模式分类器的参数,并在使用最小二乘支持向量机技术进行控制图模式识别的仿真实验研究中,分别使用遗传算法和粒子群算法对最小二乘支持向量机的参数进行了优化。以仿真实验为研究手段,对提出的 LS-SVM 控制图模式识别技术和参数优化方法进行了性能评估。

3)研究了使用 Cuscore 统计量诊断预期异常信号的相关问题

构建了使用 Cuscore 统计量诊断过程预期异常信号的单工序 ARMA 模型结构;对使用 Cuscore 统计量诊断非线性预期异常信号的性能进行了研究和评估;提出了触发 Cuscore 图的智能模型,并重点提出使用移动窗口法构建时间序列数据的正常/异常模式,研究了移动窗口的设计方法,对提出的模型进行了性能评估和仿真算例,实现了以模式识别的方式来进行变点检测,从而解决使用标准 Cuscore 统计量诊断过程预期异常信号时存在的失配问题。制造过程经过长期运行,会积累下来有关制造过程异常和故障的经验知识,通过 Cuscore 统计量来合理利用过程的先验知识,有助于提高过程异常的诊断效率。

4)研究了在多元过程中进行质量诊断的智能模型和实现方法

针对多元过程的均值矢量和协方差矩阵异常诊断问题,在研究传统 MSPC 技术的基础上,分别构建了诊断均值矢量和协方差矩阵异常的智能模型,提出了均值矢量模式和协方差矩阵模式的定义方法及其仿真数据生成方法,重点结合二元过程给出实例分析和仿真分析,对提出的诊断模型进行了性能评估,实现了在使用过程的总体统计量诊断过程异常的同时，使用 LS-SVM 模式识别技术来辨识和分离过程异常变量的功能。面对日益复杂的制造过程，研究在多变量和变量间存在相关关系的条件下，如何进行过程异常诊断和异常变量的定位，是质量诊断领域具有挑战性的任务。本书结合人工智能技术的使用，在这方面进行了有益的研究和探讨。

6.2　展望

过程质量诊断问题是一个研究范围很大的课题，由于制造过程及其异常的类型，过程异常的形成原因、机理和表现形式都是多种多样的，使得对于过程质量诊断的研究永无止境，解决该问题需要吸收并综合使用各工程学科最新的技术研究成果并应用于质量诊断领域，需要质量人员、管理人员、各学科及工程领域专家的密切配合和不懈的共同努力。本书重点探讨和研究了将人工智能技术应用于制造过程质量诊断的问题，由于该问题本身的复杂性，本人研究水平以及时间等条件的限制，在已经开展的几个方面的研究中，还有很多后续研究工作要做，作者认为主要包括以下内容：

1) 控制图模式识别方面

在使用 PNN 和 LS-SVM 技术进行控制图模式识别时，都只是针对六种基本的控制图模式进行了研究，而实际问题中存在着各种控制图异常模式的组合模式。当识别到控制图的异常模式时，有必要进一步估计过程异常的有关参数，比如，阶跃模式的幅值、循环模式的周期等参数，以上问题都尚待研究。此外，还可以考虑使用小波变换、傅里叶变换等技术进一步提高控制图的模式识别率。

2) 使用 Cuscore 统计量诊断过程中的预期异常信号方面

为了使用 Cuscore 统计量诊断过程预期异常信号，本书使用了移动窗口法和模式识别技术来检测过程观测数据的变点,以解决标准 Cuscore 统计量存在的失配问题,但这种方法在预期异常信号的变化幅值较小时效果欠佳,导致 Cuscore 统计量诊断过程的潜在效力没有得到充分发挥;没有研究和讨论当存在模型误定时 Cuscore 统计技术的稳健性问题。以上问题需要继续研究解决。

3)多元过程异常智能诊断方面

本书针对多元过程均值矢量和协方差矩阵的异常诊断问题,分别提出了基于 LS-SVM 模式识别技术的两个过程诊断模型。研究中,我们假定在诊断过程均值矢量异常变化时过程的协方差矩阵保持不变,在诊断过程协方差矩阵异常变化时过程的均值矢量保持不变,而且假定过程中变量之间的相关系数始终保持不变,不受异常因素的影响。应当进一步深入研究当上述假定条件不成立时,如何进行过程异常诊断和变量分离的问题。

4)其他方面的问题

各种智能过程诊断技术同传统 SPC 技术一样,都有其自身的优缺点。面对复杂的实际制造过程,不能指望一种诊断技术解决所有的质量诊断问题。将多种质量诊断技术结合起来组成集成化的过程质量诊断系统,并同其他的管理系统(如,计算机集成制造系统)相结合,是质量诊断的必然发展趋势。另外,质量诊断技术的研究工作应当尽量结合企业的具体实践,根据企业的实际情况,广泛使用过程数据采集软硬件技术、计算机技术来加以实现和应用,使之真正发挥提高过程质量水平、为企业创造效益的作用。

附录 以第一作者身份发表的主要学术论文

[1] Zhiqiang Cheng, Yizhong Ma. A research about pattern recognition of control chart using probability neural network [CA]. 2008 ISECS International Colloquium on Computing, Communication, Control, and Management, 2008, 2: 140-145.

（EI 收录: 20084611709509）

[2] 程志强, 马义中. 基于鲁棒 LS-SVM 的控制图模式识别 [J]. 计量学报, 2009, 3(6): 580-582.

[3] 程志强, 马义中. 基于 Cuscore 图的质量控制模式 [J]. 统计与决策, 2009, 9(2): 162-164.

（CSSCI 收录）

[4] Cheng Zhiqiang, Ma Yizhong. Control Chart Pattern Recognition Based on PSO and LS-SVM [J]. The Hong Kong Institution of Engineers Transactions, 2010, 17(1): 7-10.

（EI 收录: 20102413012049）

[5] Cheng Zhi-Qiang, Ma Yi-Zhong, Bu Jing. Mean Shifts Identification Model in Bivariate Process Based on LS-SVM Pattern Recognizer [J]. International Journal of Digital Content Technology and its Applications, 2010, 4(3): 154-170.

（EI 收录: 20104713409943）

[6] Cheng Zhi-Qiang, Ma Yi-Zhong, Bu Jing. Variance Shifts Identification Model of Bivariate Process Based on LS-SVM Pattern Recognizer [J]. Communications in Statistics—Simulation and Computation, 2011, 40(2): 286-296.

（SCI 收录:000285347600008）

[7] Zhiqiang Cheng. An intelligent method of change-point detection based on LS-SVM algorithm [J], The Hong Kong Institution of Engineers Transactions,2013,20(3):141—147.

（EI 收录:20140917370930）

A Research about Pattern Recognition of Control Chart Using Probability Neural Network

Zhiqiang Cheng

Department of Management Science and Engineering, Nanjing University of Science and Technology, Nanjing Jiangsu, 210094, China

czqme@163.com

YiZhong Ma

Department of Management Science and Engineering, Nanjing University of Science and Technology, Nanjing Jiangsu, 210094, China

yzma-2004@163.com

Abstract: In the recent years, as an alternative of the traditional process quality management methods, such as Shewhart SPC, artificial neural networks (ANN) have been widely used to recognize the abnormal pattern of control charts. But literature shows that it is difficult for a developer to select the optimum NN topology architectures in a systemic way, this kind of work was primarily done according to the developer's personal experiences and could not get desirable effect. This paper proposes to use probability neural network (PNN) to recognize the six kinds of control chart patterns (i. e. normal pattern, upward/downward mean shift pattern, upward/downward trend pattern, cyclic pattern) to improve the design effect of pattern recognition. Numerical simulation result shows that PNN has not only the feature of simpler topology structure but also the higher pattern recognition accuracy and faster recognition speed. AS the PNN pattern recognition method can get the optimum classification effect in terms of the Bayesian criterion, it is a comparable way between different manufacturing processes and suitable to be generalized as an industry criteria.

Key words: pattern recognition, probability neural network, control chart pattern

1. Introduction

[1]Statistic process control (SPC) technology, which was invented by Shewhart in 1924, has been used in the manufacturing process control widely as a monitor methodology of process quality. According to the SPC theory, there are two types of variations in the manufacturing process: one type of variation is induced by the common causes, which are unassignable, another type of variation is induced by the so-called special causes, which are assignable. The former kind of variation is inevitable and can only be diminished through management approaches but can not be deleted thoroughly from the manufacturing process; the latter kind of variation whereas can be eliminated from the manufacturing process through technique approaches (i. e., improved man, machine, material, methods, measurement, environment, which are usually abbreviated into 5M1E). The basic function of the SPC control charts is to monitor the process when the special cause variation occurs. In order to monitor and control the process when the special cause variation appears in the process, there are

---

① This work was supported by National Natural Science Foundation of China, Grant No. 70672088

six basic various SPC control chart patterns [1] (CCP, i. e. , normal, upward/downward shift, upward/downward trend, and cycle, etc. , see Figure 1) should be recognized and alarmed as soon as possible to take corresponding corrective actions. However, Shewhart SPC control charts do not always work very well in some continuous manufacturing industries, such as chemical industry, because in this type of industry the collected data from the workshop field are of the auto-correlated property which directly contradicts to the basic assumption of data in the Shewhart SPC control charts: Independence. In this kind of condition, using Shewhart control charts to monitor the process will produces more false alarm signals.

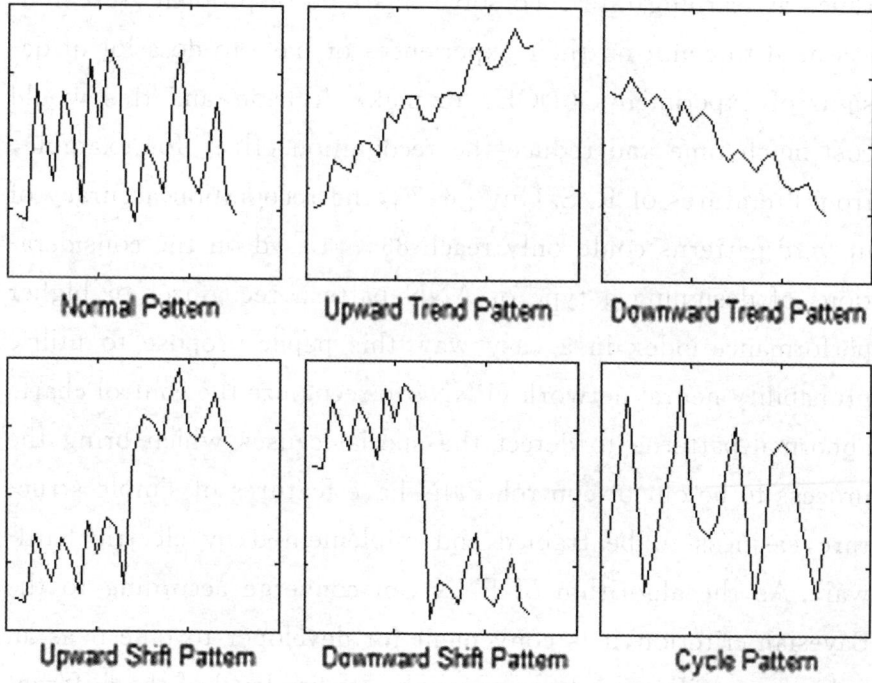

Fig 1 Six basic control chart patterns

In the recent years, as an alternative approach of traditional Shewhart SPC control charts, artificial neural networks (ANN) technology has been applied to the field of quality management to improve the performance of process. One reason to use ANN as a pattern recognizer to monitor the process is that normality of the process data is not a necessity for ANN technology. Literatures show the results of the using ANN approach for the control charts pattern recognition is inspiring [2]. According to the literatures [2,3], most of the study using NN technology to recognize control charts pattern is based on the BP[4], RBF[3] and ART neural network. Among all these NN pattern recognition methods, a common issue is the selection of the neural network structure (such as number of layers and the corresponding number of neurons in every layer) and the parameters of the NN algorithm (such as learning rate). To solve this kind of problem, researchers need to count on their experiences or have to do a lot of designs of experiments (DOE) to make decision, and this would cost much time and reduce the recognition effect, for example, from literatures of R. S. Guy [6,7], the recognition accuracy of upward patterns could only reach 86%. Based on the considerations of designing a type of ANN pattern recognizer of higher performance index in a easy way, this paper propose to utilize probability neural network (PNN) to recognize the control charts abnormal patterns to detect the special causes which bring the process to be out of control. PNN have features of simple structure, easiness to be trained and implemented by electric hardware. As the algorithm of PNN can converge according to the Bayesian criterion, it is convenient for developer to take it as an industry specification to compare the quality level of the different manufacturing processes. Numerical experimental results of using PNN to recognize control chart pattern is given out in this paper to show the advantage of this method.

2 Probability neural network model

2.1 PNN introduction

Probability neural network (PNN), which was presented by Specht in 1989 according to Bayesian classification rule and Parzen probabilistic density function, is a kind of radius basis function (RBF) neural network and is especially suitable for pattern classification. PNN has the architecture of four layers all together, where competitive output layer is an added layer on the basis of the RBF neural network which has three network layers. The names of the four PNN layers are input layer, model layer, sum layer, and competitive output layer. The structure of PNN is showed as Figure 2. The input layer has a linear transfer function; and the model layer has the same number of neurons as the input layer. Another feature of PNN structure is that the neuron numbers of the output layer is the number of pattern category of the training data set. The function of the input layer and model layer is to feed input data vector to the neural network and calculate the matching degree (with a format of distance) between input data vector and standard model data in the training data set respectively, and the calculation outcome of matching degree that is regarded as a input width parameter of activation function is send to a Gaussian type activation function (transfer function). The output of the Gaussian activation function is simultaneously the output of the model layer. The sum layer is simply used to calculate the sum of every model output. And the output layer utilizes unsupervised competitive learning rule to make decision to judge if a data set belong to a data pattern or not. Here only brief descriptions of PNN are supplied; detailed information about

PNN can be found in paper of B ERTHOLDM and D IAMONDJ [8].

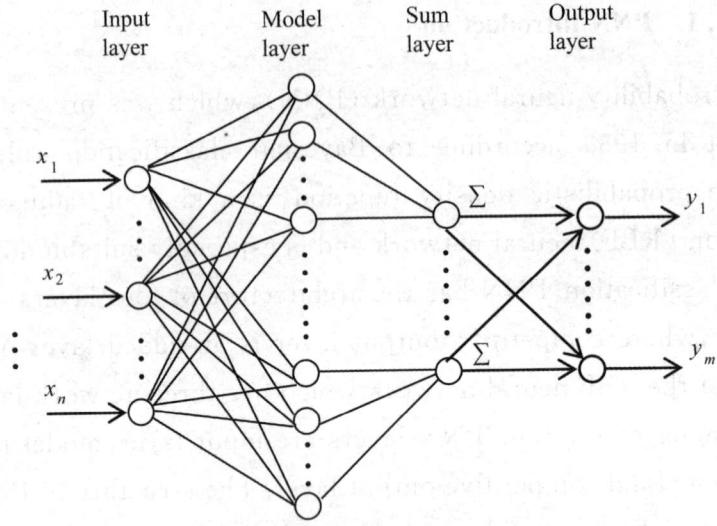

Fig 2　Hierarchical structure of PNN

As far as the performance of PNN is concerned, PNN is a NN algorithm in terms of the statistical Bayesian minimum-risk principle, and is equal to a Bayesian optimal classification recognizer without local minima, so it can almost get the best classification effect provided enough training data are fed to it. Meanwhile, PNN use fully feed-forward parallel algorithm without back-propagation calculation, so only one network layer need to be trained and it has faster training speed than BP neural network does. Another advantage of PNN is the easiness to implement in technique.

2.2　PNN design in this paper

As for the architecture of the PNN in this paper, the number of neurons in the input layer is designed to the dimension of the input data, the number of neurons of the output layer is the pattern category number of the training data set, and there is no need to design the hidden layer (model layer and sum layer) through DOE or personal experience. Undoubtedly, those features could greatly simplify the design of the NN comparing to the design of BPN hidden layer [4,6,7]. In this paper, the input layer of the PNN model contains 25 neurons used for 25 consecutive points in a SPC control charts, and this is in conformity to the AIAG (Automotive Industry Action Group) SPC criteria. The output layer consists of six neurons which are corresponding to the six types of control chart patterns proposed in this research respectively. Only one parameter need to be designed is the width of the Gaussian transfer function. The designed PNN architecture diagram using MATLAB software tool, with 25 input neurons and 6 output neurons, is simply presented as Figure 3, where shows that the transfer function of PNN is radial basis Gaussian function, the width of the Gaussian transfer function could been set through parameter SPREAD of MATLAB function "newpnn" in the simulation experiment.

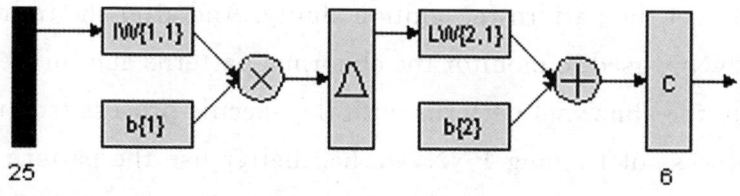

Fig 3 PNN design

In this paper, number of neurons of the input layer is actually referred to the window size in on-line SPC control chart. The size of this window can influence the recognition ability of this PNN recognition model greatly. A small window size will typically detect special cause patterns more quickly, while it may also result in a short in-control ARL (Average Run Length, equivalent to a high Type I error) when process is in control; a big window size can lower the recognition efficiency by increasing the time required to detect patterns (equivalent to a higher Type II error, or longer out-of-control ARL, which is preferred to be short when the process is out of control). Another aim of this research is trying to give out a comparable pattern recognition approach for the process customers to evaluate the quality performance of the manufacturing process when the NN approach is used in different manufacturing factory, so we take the automobile industry standard from AIAG SPC handbook as a reference to select the SPC window size as 25 points to design PNN pattern recognizer.

3 Training data set preparation

3.1 Data sets generation

Before the PNN is used to work as a control chart pattern (CCP) recognizer, it must be trained through enough numbers of data to get the pattern recognition ability. And after the training, the PNN is used to monitor the abnormal patterns and further associate the abnormal patterns with the specific process factors. In the process of training PNN, we had better use the pattern data (natural and unnatural) collected from a real process to train and test the PNN, but it is difficult and costly for us to collect the large amount of various pattern data. To economically generate the data to train and test the PNN used for control charts pattern recognition, this paper use Monte-Carlo simulation method to produce the required data sets. In this research, the six different pattern data sets for training and testing are simulated using the following equations:

For data set $x(t)$,
$$x(t) = \mu + d(t) + r(t) \tag{1}$$
where, t is time of sampling, $x(t)$ is the sample value at time t, μ is mean value of the normal process variable being monitored, $r(t)$ is the common cause variation at time t and follow a normal distribution with zero mean and standard deviation σ, $d(t)$ is the special cause disturbance at time t ($d(t)$ is zero when only natural pattern data are existing). Without loss of generalization, in this paper, $\mu=0$, $\sigma=1$. The task of pattern recognition is to identify the change status of $d(t)$.

1) For normal pattern,
$$x(t) = r(t) \tag{2}$$
Where, $r(t) \sim N(0,1)$, $d(t)=0$.

2) For upward and downward trend pattern,
$$x(t) = \pm k \times t + r(t) \tag{3}$$
Where: k is trend slope in terms of σ, $0.1\sigma \leqslant |k| \leqslant 0.3\sigma$, σ is an random value in this paper.

3) For upward and downward shift pattern,
$$x(t) = \pm 1(t-t_0) \times s + r(t) \tag{4}$$
Where, $1(t-t_0)=1$, when $t \geqslant t_0$, and $1(t-t_0)=0$, when $t < t_0$, and t_0 is a parameter to determine the shift position ($t_0=12$, 13 or 14, random integer value in this paper). And s is the shift magnitude in terms of σ, and $\sigma \leqslant |s| \leqslant 3\sigma$ in this paper, random value.

4) For cyclic pattern,
$$x(t) = A \times \sin(2\pi t/T) + r(t), \tag{5}$$
Where, A is cycle amplitude in terms of σ, $\sigma \leqslant A \leqslant 3\sigma$, in this paper, random value. T is the cycle period ($2 \leqslant T \leqslant 8$), and $T=8$ in this paper.

3.2 Data sets preparation

Before the data sets being inputted to PNN, data sets need standardization and coding.

1) Standardization

Standardization here is to scale the train and test data into a constant range (e.g. $[-1,1]$ or $[-5,5]$), it is because that a trained NN can only accept the input data within a certain range. If a neural network is trained with a range of data to get the trained working ability, and latter on it is tested or used with another range of data, the neural network can not work well with the trained working ability. The following formula is used to standardize the data

$$z(t) = (x(t) - \mu)/\sigma \qquad (6)$$

Where $z(t)$ is the standardized value from $x(t)$.

In an actual application, the equation to standardize the data is shown below

$$z(t) = (x(t) - E(\mu))/E(\sigma) \qquad (7)$$

Where $E(\mu)$ and $E(\sigma)$ are the estimated values from the sampling data which are collected from the manufacturing process.

2) Coding

After data standardization, coding approach is used to filter the impact of small noise signal coming from the natural cause variations and to facilitate the convergence of the neural network training process (that also means coding can reduce the time to train a neural network). In the process of data coding, the main features of the data must be kept retained, in other words, if the original analog data are coded into digital data with too few rates, the features of the original analog data can not be retained; Whereas, if the original analog data are coded into digital data with too much rates, it would take too much time to process them in the later calculation and reduce the real-time ability of the PNN recognizer. The coding principle and its formula in this paper is listed as below

$$\begin{cases} y(t)=25, x(t) \geqslant 4.9 \\ y(t)=-25+N, -4.9+0.2(N-1) \leqslant x(t) < -4.9+0.2N \\ y(t)=-25, x(t) < -4.9 \end{cases}$$

(8)

Where N is a parameter to determine the coding rate (Here, $N=1,2,3,\cdots,49$), and $y(t)$ is the coded values from $x(t)$.

4 Experiment results using PNN recognizer

4.1 Experiment simulation data group

In this study, all together 60000 (25 * 400 * 6) pattern data were used for training and testing the designed PNN, and there into 45000 (25 * 300 * 6) pattern data were used to train the PNN, and the other is used to test the PNN. Every pattern had the same training data to make the PNN neural network get the same training condition for the six different abnormal patterns.

4.2 Experiment parameter design

In the process of simulation experiment through MATLAB software, we could find that the simulation parameter SPREAD of MATLAB function "newpnn", which is used to create a PNN, had a great impact on the simulation outcomes. When the experiment was been doing, the parameter SPREAD, which represented the width of radius basis Gaussian transfer function, was designed to increase from 0.5 to 16 with an increment of 0.5, 1 or 2, and it was found that the simulation experiment could get better recognition accuracy when SPREAD has a value of 9, and the final result of this research was made under this parameter setting. The following Table 1 recorded the effects of the different parameter SPREAD setting. The width of the radius basis function was the only parameter that had to design in this research.

5 Performance evaluation

5.1 Recognition accuracy

This paper takes recognition accuracy (RA) as the performance index to evaluate the recognition ability of the trained PNN recognizer. The test outcomes about RA of the six patterns are listed in Table 2. The RA percentage of every pattern is the mean of one thousand times experiments.

From data of Table 2 and the comparison with the results data from R. S. Guh [6,7], it can see that PNN recognizer obviously worked better than research [6,7] on the six basic pattern with percentage RA all above 0.93. However, it is also apparent to see that with the same training samples and parameters setting, PNN recognizer did not have just the same recognition ability for every pattern. This condition depends on whether the features of the six patterns are distinct from each other. From further observation of the experiment outcome data, it could be found the reason why the experiment could not get the approximate 100% RA about the upward/downward trend pattern and upward/downward shift pattern was that the PNN pattern recognizer mistake upward/downward trend pattern for upward/downward shift pattern respectively, and vice versa. So this paper suggests integrating PNN pattern recognizer with other appropriate technique (e.g. wavelet analysis technique) to distinguish shift pattern from trend pattern in order to get better recognition effect.

5.2 ARL calculation

This research also adopts average run length (ARL) as performance index to evaluate the recognition speed of the trained PNN. ARL means the average number of the samples before an out-of control signal is produced by the pattern recognizer. Usually out-of-control ARL is expected to be small whereas in-control ARL is expected to be big. This paper calculated the ARL using a moving window (with fixed window size 25) methodology to calculate the ARL, the basic explanation of how the ARL was computed by PNN recognizer is showed in Figure 4.

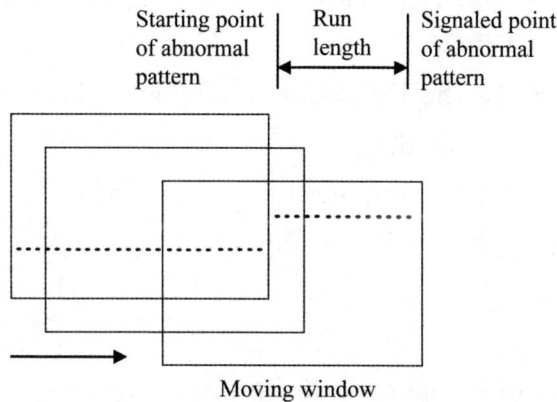

Fig 4　Computation of run length

At the beginning of computation run length, a sequence of 25 normal pattern simulated data in the moving window was initially assumed to be offered to the PNN recognizer, then the moving window moved forward and add one datum from a kind of abnormal pattern simulated datum into the moving window and subsequently fed them to the PNN recognizer to analyze if this kind of abnormal pattern occurred. Every abnormal pattern data here were generated with the same way as the training data. With the window moving and the PNN recognition process going by, abnormal alarm information would be signaled out by the PNN recognizer, and the number between the start sampling point of an abnormal pattern and the alarm point of an abnormal pattern was the expected run length. And to get ARL of every abnormal pattern, the simulation experiment would be done 10000 times for each abnormal pattern and calculate the mean of 10000 run length of every abnormal pattern. From the description of the ARL calculation, it could be found that moving window method for ARL calculation was a very similar process to the on-line product quality monitoring process. Detailed procedure to calculate ARL of a certain kind of abnormal pattern by moving window method is summarized as below,

1) Determine the threshold value θ for a kind of abnormal pattern to be recognized.

2) Set the initial sampling time t to 25.

3) Take the recent simulated data
$X_j(j = t-24, t-23, \cdots, t-1, t)$ from the abnormal pattern data set.

4) Standardize and code X_j into $\overline{X}_j$.

5) Build up data $\overline{X}_j$ into a vector V_i,
Where $V_i = (\overline{X}_{t-24}, \overline{X}_{t-23}, \cdots, \overline{X}_{t-1}, \overline{X}_t)$, and feed V_i to the PNN recognizer to get the output O.

6) If $O = \theta$, then get the conclusion that the abnormal pattern has been detected, and go to step 7); otherwise increment t by 1, and go to step 3) until this kind of abnormal pattern was signaled.

7) Calculate and register the run length of this experiment.

The ARL of the researched abnormal pattern recognition is the average of run length of 10000 times experiments.

According to the ARL calculation procedure introduced above, simulation experiments were done in this research. Numerical results of ARL calculation for the five kind of abnormal patterns were listed in Table 3.

From the experiment results of the ARL calculation, it can be concluded that PNN pattern recognizer has a faster recognition speed than the traditional Shewhart control chart.

6 Conclusion

Based on the consideration of put forward an agile and comparable way to design ANN control chart pattern recognizer of high performance to fulfill the demand of agile manufacturing process, a PNN pattern recognizer is proposed in this paper. Experimental results show that PNN control chart pattern recognizer performs well on six typical control chart patterns according to the two performance index, recognition accuracy (RA) and ARL. As it is easy to realize PNN pattern recognizer in the engineering field (both software and hardware), this research suggests the feasibility to generalize the PNN control chart pattern recognizer as a comparable approach in different manufacturing industry to monitor the process abnormality. The easy design of PNN pattern recognizer and the experiment results show that the PNN pattern recognizer is efficient and suitable for manufacturing process to improve the efficiency of product implement process and reduce product lead-time.

Tab 1 Width design of radius basis transfer function parameter SPREAD

SPREAD \ RA	RA (Normal)	RA (Upward Trend)	RA (Downward Trend)	RA (Upward Shift)	RA (Downward Pattern)	RA (Cycle)
0.5	0.999	0.031	0.029	0.001	0.001	0.001
1	0.949	0.937	0.933	0.901	0.902	0.963
1.5	0.940	0.933	0.933	0.902	0.891	0.961
2	0.946	0.931	0.931	0.898	0.897	0.961
4	0.955	0.936	0.935	0.909	0.900	0.965
6	0.976	0.935	0.938	0.924	0.923	0.966
8	0.989	0.940	0.939	0.942	0.941	0.969
8.5	0.993	0.937	0.943	0.946	0.945	0.967
9	0.995	0.940	0.936	0.946	0.951	0.967
9.5	0.995	0.932	0.932	0.946	0.9488	0.964
10	0.996	0.932	0.930	0.946	0.943	0.957
12	0.999	0.919	0.912	0.943	0.942	0.954
14	0.999	0.900	0.900	0.937	0.934	0.950
15	0.998	0.889	0.890	0.929	0.934	0.947
16	0.999	0.877	0.879	0.929	0.930	0.948

Tab 2 Experiment result of recognition accuracy

Pattern	Normal	Upward Trend	Downward Trend	Upward Shift	Downward Pattern	Cycle
RA of PNN	0.995	0.940	0.936	0.946	0.951	0.967
RA from R. S. Guh. [6,7]		0.86/0.86	0.86/0.89	0.90/0.90	0.90/0.89	0.92/0.94

Tab 3　Results of ARL calculation

Pattern type	Upward trend	Downward trend	Upward shift	Downward shift	Cycle
ARL	20.62	20.34	20.11	19.78	16.56
σ of ARL	3.27	3.30	3.84	2.82	7.29

References

[1] Wimalin Sukthomya, James Tannock, "The training of neural networks to model manufacturing processes", Journal of Intelligence Manufacturing, 2005, Vol. 16, pp. 39—51.

[2] Ruey-Shy Guh, Yi-Chih Hsieh. "A neural network based model for abnormal pattern recognition of control charts", Computer & Industrial Engineering, 1999, Vol. 36, pp. 97—108.

[3] Deborah F. Cook, Chih-Chou Chiu, "Using radial basis function neural networks to recognize shifts in correlated manufacturing process parameters", IIE Transactions, 1998, Vol. 30, pp. 227—234.

[4] Hwarng HB, Hubele NF, "Back-propagation pattern recognizers for X-bar control charts: methodology and performance", Computer & Industrial Engineering, 1993, Vol. 24, pp. 219—235.

[5] Susanta Kumar, "A study on the various features for effective control chart pattern recognition", International Journal of Advanced Manufacturing Technology, 2007, Vol. 34, pp. 385—398.

[6] Ruey-Shiang Guh, "Robustness of the neural network based control chart pattern recognition system to non-normality", International Journal of Quality & Reliability Management, 2002, Vol. 19(1), pp. 97—112.

[7] R.-S. Guh, F. Zorriassatine, J. D. T. Tannock, and C.

O'Brien,"On-line control chart pattern detection and discrimination—a neural network approach",Artificial Intelligence in Engineering,1999,Vol. 13,pp. 413—425.

[8] B ERTHOLDM,D IAMONDJ,"Constructive training of probabilistic neural networks",Neurocomputing,1998,Vol. 19(1—3),pp. 167—183.

[9] A. S. ANAGUN,"A neural network applied to pattern recognition in statistical process control",Computer ind. Engng,1998,Vol. 35,Nos 1—2,pp. 85—188.

◀ 附录　以第一作者身份发表的主要学术论文

基于鲁棒 LS-SVM 的控制图模式识别

程志强，马义中

南京理工大学 经济管理学院，江苏南京，210094

摘要：本文提出基于鲁棒最小二乘支持向量机的控制图模式识别方法，并研究其应用于过程质量诊断的可行性、有效性。理论研究和仿真试验结果表明，该方法对于标准的六种控制图模式都具有很高的模式识别率，训练模式识别器所需样本少且训练结果泛化能力强，计算方法简单迅速。以上特点，对于降低质量诊断成本，以及该技术在过程现场的应用具有现实意义。

关键词：控制图，最小二乘支持向量机，模式识别

中图分类号：F253.3，TP391.4　　　　**文献标示码**：A

Control Chart Pattern Recognition Based on Robust LS-SVM

Cheng Zhi-Qiang, Ma Yi-Zhong

School of Economics and Management, Nanjing University of Science and Technology, Nanjing Jiangsu, 210094

Abstract: This paper propose an approach based on the robust least squares support vector machines (RLS-SVM) for con-

---

① 金项目：国家自然科学基金资助项目(70672088)
作者简介：程志强(1968—)，男，河南安阳人，南京理工大学博士生，从事质量工程技术领域的研究；马义中(1964—)，男，河南泌阳人，南京理工大学教授、博士生导师，从事质量工程与管理领域的研究。

trol charts pattern recognition, the applied feasibility and validity of this technique in process quality diagnosis is also investigated. Theoretical research and experimental results show that this approach performs well upon the six typical control charts pattern recognition with high recognition accuracy, simple computation and fast training process, and the preeminent generalization ability on the condition of small sample size. All those performances have significant meaning for reducing the cost of quality diagnosis and the application of this technique in the manufacturing process field.

Key words: control chart, least squares support vector machines, pattern recognition

1 引言

对控制图正常模式和各种异常模式进行有效识别,是进行质量诊断与监控的重要手段。目前对于控制图模式的识别大量地是采用基于经验风险最小化(Empirical Risk Minimization, ERM)的人工神经网络(Artificial Neural Network,ANN)方法或以神经网络为主同时结合其他技术(如,小波)的方法[1,2,3]。使用神经网络方法进行控制图模式识别,存在大样本训练,训练时间长、训练结果泛化能力差、总体识别率不高等缺陷,基于质量成本考虑,上述方法难于推广应用。基于统计学习理论(Statistical Learning Theory,SLT)[4]的支持向量机(Support Vector Machine,SVM)方法采用结构风险最小化(Structural Risk Minimization)原理来训练样本误差,其模型结构由支持向量决定,使得SVM可以在小样本情况下对控制图模式进行迅速有效识别,并且训练结果泛化能力强,符合质量管理中减少观测成本和敏捷制造的需要。最小二乘支持向量机(Least Squares SVM,LS-SVM)[5,6,7]将标准SVM中的二次规划问题转化为线性方程组求解,算法简单的运算速度快,但LS-SVM使用平方和作为性

◀ 附录 以第一作者身份发表的主要学术论文

能评价指标,在过程出现异常观测值或者过程误差变量的高斯分布假设不成立时(如,大量存在的化工连续生产过程)会降低结论的鲁棒性,为此本文提出使用鲁棒统计方法加权改进 LS-SVM 的鲁棒性,以实现过程在小样本情况下六种典型控制图模式的快速准确识别。仿真试验结果证明了该方法在控制图模式识别中具有识别率高、识别速度快的良好性能。

2 最小二乘支持向量机

设训练样本数据集为 $\{x_i, y_i\}_{i=1}^{N}$,N 为样本容量,输入数据 $x_i \in R^n$,输出数据 $y_i \in R$。SVM 的思想是通过非线性映射 $\phi(\cdot)$ 将样本映射到高维特征空间 $F(\phi:R^n \rightarrow F)$,并在 F 中构造线性回归函数:

$$y_i = \omega^T \phi(x_i) + b \tag{1}$$

LS-SVM 使用训练集错分误差的二范数 $\sum_{i}^{N} \xi_i^2$ 代替标准 SVM 中的错分误差之和 $\sum_{i}^{N} \xi_i$,形成下面的优化问题:

$$\begin{cases} \min \quad J(\omega, \xi) = \frac{1}{2}\omega^T\omega + \frac{1}{2}\gamma \sum_{i=1}^{N} \xi^2 \\ \text{s.t.} \quad y_i = \omega^T \phi(x_i) + b + \xi_i \end{cases} \tag{2}$$

式中,$i = 1, \cdots, N$,ξ_i 为错分随机误差,$\gamma > 0$ 为对错分误差平方和的惩罚系数,它折中经验误差和模型的复杂程度以使所求 SVM 具有较好的泛化能力,且当 γ 增大时,经验风险小。构造拉格朗日函数:

$$L(\omega, \xi, \alpha, b) = J(\omega, \xi) - \sum_{i=1}^{N} \alpha_i (\omega^T \phi(x_i) + b + \xi_i - y_i) \tag{3}$$

其中,$\alpha = [\alpha_1, \alpha_2, \cdots, \alpha_N]^T$,$i = 1, \cdots, N$,$\alpha$ 为拉格朗日乘子,根据 KKT 优化条件,对式(3)分别对 ω, ξ, α, b 求偏导并置零,得到:

$$\begin{cases} \omega = \sum_{i=1}^{N} \alpha_i \phi(x_i) \\ \alpha_i = \gamma \xi_i, i=1,2,\cdots,N \\ y_i = \omega^T \phi(x_i) + b + \xi_i, i=1,2,\cdots,N \\ \sum_{i=1}^{N} \alpha_i = 0 \end{cases} \quad (4)$$

消去 ω 和 ξ 后,优化问题变为求解如下线性方程:

$$\begin{bmatrix} \Omega + \gamma^{-1}I & l_N \\ l_N^T & 0 \end{bmatrix} \begin{bmatrix} \alpha \\ b \end{bmatrix} = \begin{bmatrix} y \\ 0 \end{bmatrix} \quad (5)$$

其中,$l_N = [1,1,\cdots,1]^T$,$\alpha = [\alpha_1,\alpha_2,\cdots,\alpha_N]^T$,$y = [y_1,y_2,\cdots,y_N]^T$,$\Omega$ 为一个方阵,其元素 $\Omega_{ij} = \psi(x_i,y_j) = \phi(x_i)^T\phi(x_i)$,$i,j = 1,\cdots,N$,$\psi(\cdot)$ 定义为满足 Mercer 条件的核函数。解方程(5)求出 α 和 b 后,可得如线性回归模型

$$y(x) = \sum_{i=1}^{N} \alpha_i \psi(x,x_i) + b \quad (6)$$

本文中 ψ 取 RBF 核函数:

$$\psi(x,y) = \exp(\frac{-\|x-y\|^2}{2\sigma^2}) \quad (7)$$

参数 γ 和 σ 的取值对 LS-SVM 的函数估计精度和收敛速度会产生明显影响。

3 最小二乘支持向量机的加权改进

由于 LS-SVM 采用 $\sum_{i}^{N} \xi_i^2$ 作为评价函数,当训练数据中存在异常值或非高斯噪声时,对分类函数的估计可能失去稳健性。而对于异常数据点的正确分类及在非高斯噪声情况下对质量控制图的正确识别正是本文研究的内容。本文采用鲁棒统计中加权的方法[8]使得最小二乘向量机来获得解的鲁棒性,为此对式(4)中的松弛误差变量 $\alpha_i = \gamma \xi_i$ 引入加权因子 v_i,可得下面的最优分类超平面问题:

$$\begin{cases} \min\ J(\omega^*,\xi^*) = \frac{1}{2}\omega^{*\mathrm{T}}\omega^* + \frac{1}{2}\gamma\sum_{i=1}^{N}\upsilon_i\xi_i^2 \\ \mathrm{s.t.}\ \ y_i = \omega^{*\mathrm{T}}\phi(x_i) + b^* + \xi_i^* \end{cases} \quad (8)$$

相应的拉格朗日函数为:

$$L(\omega^*,\xi^*,\alpha^*,b^*) = J(\omega^*,\xi^*) - \sum_{i=1}^{N}\alpha_i^*(\omega^{*\mathrm{T}}\phi(x_i) + b^* + \xi_i^* - y_i)$$

(9)

式中"*"表示待定未知变量。根据KKT条件并消去ω^*,ξ^*,可得如下线性方程:

$$\begin{bmatrix} 0 & l_N^{\mathrm{T}} \\ l_N & \Omega + V_\gamma \end{bmatrix}\begin{bmatrix} b^* \\ \alpha^* \end{bmatrix} = \begin{bmatrix} 0 \\ y \end{bmatrix} \quad (10)$$

其中,$V_\gamma = \mathrm{diag}[1/\upsilon_1, 1/\upsilon_2, \cdots, 1/\upsilon_N]$,加权因子$\upsilon_i$是基于误差$\xi_i$的取值,可以通过下式得到:

$$\upsilon_i = \begin{cases} 1, & \text{when}\ \ |\xi_i/\hat{s}| \leqslant c_1 \\ \dfrac{c_2 - |\xi_i/\hat{s}|}{c_2 - c_1} & \text{when}\ \ c_1 \leqslant |\xi_i/\hat{s}| \leqslant c_2 \\ 10^{-4} & \text{others} \end{cases} \quad (11)$$

其中,ξ_i由式(4)求出。c_1,c_2为常数,经验典型取值为$c_1 = 2.5$,$c_2 = 3$,$\hat{s}$是误差变量ξ_i的鲁棒估计值[8],

$$\hat{s} = 1.483 MAD(\xi_i) \quad (12)$$

MAD为中值绝对偏差。通过上述对于误差项ξ_i的处理,使得当ξ_i中存在异常值或不满足高斯特性时,获得模型参数的鲁棒估计。由式(10)求得b^*和α_i^*后,且令

$$f(x) = \sum_{i=1}^{N} y_i\alpha_i^*\psi(x,x_i) + b^* \quad (13)$$

$f(x)$是由核$\psi(x,z)$隐式定义的特征空间中的最大间隔超平面,决策规则由$\mathrm{sgn}[f(x)]$给出。其中,$\mathrm{sgn}(x) = \begin{cases} 1, x > 0 \\ -1, x < 0 \end{cases}$。由于本文使用RBF核函数,$b^*$已经包含在$\psi(x,x_i)$中[9],因而决策函数变为:

$$f(x) = \sum_{i=1}^{N} y_i \alpha_i^* \psi(x, x_i) \qquad (14)$$

4 控制图模式数据及分类方案

4.1 质量控制图的数据描述

本文对六种基本质量控制图模式[10]进行分类识别,各类控制图数据集 $x(t)$ 的仿真描述为:

$$x(t) = \mu + d(t) + r(t) \qquad (15)$$

$x(t)$ 是在 t 时刻的控制图数据采样值;μ 是过程受控时的过程变量均值;$r(t)$ 是在 t 时刻的过程的偶然随机干扰成分,一般 $r(t)$ 可以假定是方差为 σ^2、均值 μ 为零的高斯白噪声序列;$d(t)$ 是 t 时刻的特殊原因干扰值。实际的应用中,μ 和 σ^2 使用过程变量采样数据的估计值。不失一般性,本文仿真时 $\mu=0, \sigma=1$。

控制图的模式由 $d(t)$ 决定:

1) 正常模式

$$x(t) = r(t), d(t) = 0 \qquad (16)$$

2) 上升/下降趋势模式

$$x(t) = k*t + r(t), d(t) = k*t \qquad (17)$$

式中,k 是上升下降趋势的斜率,$0.15\sigma \leqslant |k| \leqslant 0.3\sigma$。$k>0$ 时为上升趋势,$k<0$ 时为下降趋势。

3) 向上/向下阶跃模式:

$$x(t) = 1(t-t_0)*s + r(t), d(t) = 1(t-t_0)*s \qquad (18)$$

式中,当 $t \geqslant t_0$ 时,$1(t-t_0)=1$;当 $t < t_0$ 时,$1(t-t_0)=0$。t_0 决定阶跃发生的时刻。s 是阶跃的幅度,$1.5\sigma \leqslant |s| \leqslant 3\sigma$,$s>0$ 为向上阶跃,$s<0$ 为向下阶跃。

4) 周期循环模式:

$$x(t) = d(t) + r(t), d(t) = A*\sin(2\pi t/T) \qquad (19)$$

式中,A 是幅值,$\sigma \leqslant A \leqslant 3\sigma$;$T$ 为周期,$2 \leqslant T \leqslant 8$,本文仿真时取 $T=7$。

为了消除量纲和噪声对于训练和测试的影响,仿真数据以及过程现场的观测数据需要进行标准化和编码才能用于 SVM 的训练和测试。对于本文仿真数据使用的标准化和编码公式如下:

$$z(t) = ((x(t) - E(\mu))/E(\sigma)) \qquad (20)$$

$$\begin{cases} y(t) = 25, z(t) \geqslant 4.9 \\ y(t) = -25 + N, -4.9 + 0.2(N-1) \leqslant z(t) < -4.9 + 0.2N \\ y(t) = -25, z(t) < -4.9 \end{cases}$$

$$(21)$$

其中,$N = 1, 2, 3, \cdots, 49, E(\mu)$ 和 $E(\sigma)$ 分别是由现场观测数据得到的过程均值和方差的估计。对于本文仿真数据,$E(\mu) = \mu$,$E(\sigma) = \sigma$。$y(t)$ 经过标准化和编码处理后得到数值。

4.2 模式分类方案设计

SVM 建立的是二类模式分类器,对于多类分类识别问题,可以通过具有较高训练精度和泛化能力的一对一(one against one)方案建立多类分类器。对于 C 类分类问题,该方案在每两个类别的数据之间训练出一个分类器,最终转化为建立 $C(C-1)/2$ 个两类分类器问题。对于一个质量控制图测试样本 x,该方法依次将 x 输入 $C(C-1)/2$ 个两类分类器,然后采用多者为胜的投票法进行分类。当使用对第 m 和第 n 类二分类器 ($m \neq n, m, n \leqslant C$) 判决时,如果该分类器判决 x 属于第 m 类,就给第 n 类得票数加 1,否则就给第 j 类得票加 1。$C(C-1)/2$ 个二分类器都对该样本进行测试,最后样本属于得票数最多的那一类。即,通过对第 m 类和第 n 类数据训练得到二分类函数 $f_{mn}(x)$,对于数据 x 计算下面的函数:

$$f_m(x) = \sum_{m \neq n, m=1}^{C} sgn(f_{mn}(x)) \qquad (22)$$

则数据 x 被确定属于下面的一类,

$$\underset{m=1,\cdots,C}{\arg\max} f_m(x) \qquad (23)$$

对于本文的六类控制图模式,一对一分类方案不会导致运算量过大问题。

5 仿真试验结果

本文控制图模式识别仿真试验的过程用图 1 表示如下，

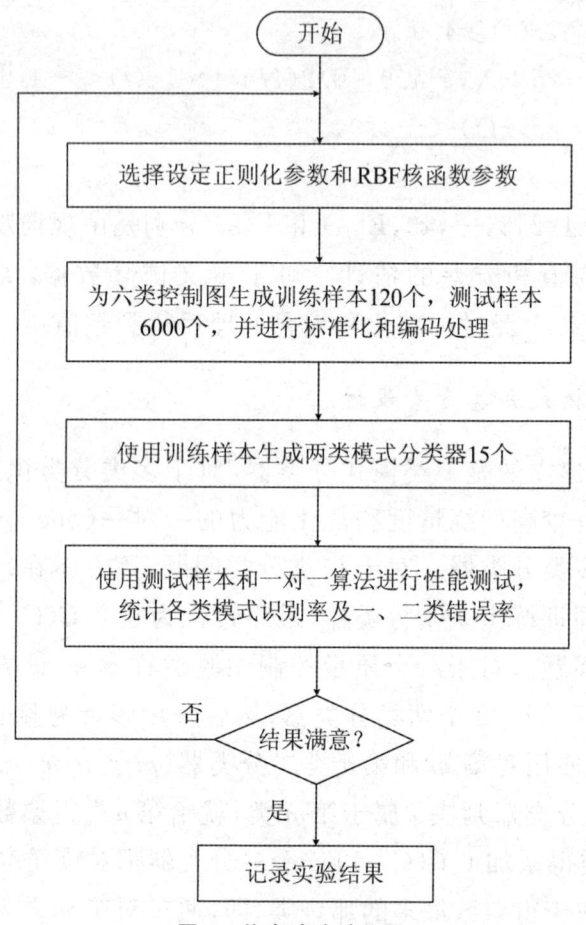

图 1　仿真试验过程图

根据上述步骤，本文在训练二分类回归函数 $f_{mn}(x)$ 时对两类控制图各使用训练样本 20 个，测试样本对于各类控制图模式都是 1000 个，6 类控制图分别标记为 $y_i = i, i = 1,2,3,4,5,6$，且通过手动交叉验证试验得到正规化参数与核函数参数的取值 $\gamma = 10, \sigma = 2.2$，仿真试验得到各类控制图的识别率，一类错误率，二类错误率，如表 1 所示。

表 1　鲁棒 LS-SVM 控制图模式识别率(%)

六个控制图模式识别率						一类错误率	二类错误率	总识别率
正常模式	向上趋势	向下趋势	向上阶跃	向下阶跃	周期模式			
97.2	99.0	99.5	96.2	93.3	97.4	2.0	0.88	97.1

从试验结果可以看出，鲁棒 LS-VCM 对于控制图具有很高的模式识别率。试验中使用的测试样本数量远远大于训练样本数量，由此可以验证使用该方法进行控制图识别的泛化能力和对于奇异控制图分类的鲁棒性。本研究使用的训练样本只有 $6×20=120$ 个，比通常的神经网络方法上万个训练数据而言大大减少而且训练速度快。以上试验结果并非在参数 γ,σ 的最优取值下获得，现场应用时如果采用诸如遗传算法进行优化，可以得到更好的识别效果。

6　结论

(1) 控制图的模式识别是质量诊断与改进的基础。本文提出使用基于鲁棒 LS-VSM 方法进行质量控制图模式识别，避免了传统神经网络模式识别方法经验风险最小化，结构选择困难等缺点。

(2) 仿真试验结果表明，该方法具有小样本训练情况下识别率高，计算速度快等优点，尤其是该方法在小训练样本情况下表现出来的超强泛化能力以及可以应用于非高斯特性制造过程的特点，对于降低质量诊断成本，提高质量诊断速度，以及该技术在过程现场的推广具有现实意义。

参考文献

[1] Ruey-Shiang Guh. Robustness of the neural network based control chart pattern recognition system to non-normality.

International Journal of Quality & Reliability Management, 2002,19(1):97—112.

[2] Ruey-shiang,Guh. A hybrid Learning-based Model for On-line Detection and Analysis of Control Chart Patterns. Computers & Industrial Engineering,2005,49:35—62.

[3] Yousef Al-Assaf. Recognition of control chart patterns using multi-resolution wavelets analysis and neural networks. Computers & Industrial Engineering,2004,47:17—29.

[4] Vapnik V N. The Nature of Statistical Learning Theory [M]. New York:Spring Verlag,1999.

J. A. K. Suykens?,J. De Brabanter,L. Lukas,J. Vandewalle.

[5] J. A. K. Suykens,J. Vandewalle. Least squares support vector machine classifiers. Neural Processing Letters,1999,9(3):293—300.

[6] Gavin C. Cawley, Nicola L. C. Talbot. Improved sparse least-squares support vector machines. Neurocomputing Letters, 2002,48:1025—1031.

[7] J. A. K. Suykens. Sparse least squares support vector machine classifiers. in:Proceedings of the European Symposium on Artificial Neural Networks (ESANN-2000) Bruges,Belgium, 26—28 April,2000,pp. 37—42.

[8] F. R. Hampel,E. M. Ronchetti,P. J. Rousseeuw,W. A. Stahel, Robust Statistics: The Approach Based on Influence Functions,Wiley,New York,1986.

[9] C. L. Huang,C. J. Wang. A GA-based feature selection and parameters optimization for support vector machines. Expert Systems with Applications,2006,31:231—240.

[10] Wimalin Sukthomya,James Tannock. The training of neural networks to model manufacturing processes. Journal of Intelligence Manufacturing,2005,16:39—51.

基于 Cuscore 图的质量控制模式研究

程志强,马义中

南京理工大学 经济管理学院,江苏南京,210094

摘要:不断提高过程质量诊断及改进的效率,是过程质量管理的追求目标。本文提出不仅使用传统休哈特控制图,而且使用 Cuscore 图对工业过程实施监控,达到有效利用以往的过程异常知识和处理经验、提高改进效率的目的。研究和仿真结果表明,Cuscore 图在有色噪声过程环境中对于预期微小异常信号的诊断比传统控制图更加迅速有效,将 Cuscore 图整合到已有的质量管理模式中,可以提高过程质量诊断与异常处理的效率。

关键词:统计过程控制,控制图,Cuscore 统计量,六西格玛管理

中图分类号:F253.3 **文献标示码**:A

Research about Quality Control Model Based on Cuscore Charts

Cheng Zhi-Qiang, Ma Yi-Zhong

School of Economics and Management, Nanjing University of Science and Technology, Nanjing Jiangsu, 210094

Abstract: To continually improve the efficiency of process quality diagnosis and improvement, is the aim of process quality management. For the sake of using the former accumulated process abnormity knowledge and treatment experience, this research proposes to use both traditional Shewhart control charts and Cuscore control chart to monitor the industrial process to advance the improvement efficiency. Research and simulation results showed that Cuscore control chart could detect the anticipated small abnormal signal more effectively than the traditional control charts could do in a colored-noise process condition. The integration of Cuscore chart with traditional quality management model can greatly improve the efficiency of process diagnosis and abnormity treatment.

Key words：Statistical Process Control，Control Chart，Cuscore Statistic，Six Sigma Management

1 引言

使用传统统计过程控制（Statistical Process Control，SPC）控制图（休哈特，Cusum，EWMA）等质量工具，可以通过对于观测数据的处理检测到制造过程中存在异常原因并给出报警，但却无法给出异常的性质、大小、部位，以及造成异常的具体物理原因和处理方法等方面的信息。传统 SPC 技术需要假定过程噪声是白色的。对于一个制造企业，伴随其发展，会根据以往的操作经验不断积累关于过程故障信号的知识（如，函数表达式、类型、大小等），并且逐渐积累形成与其相对应的故障原因和排除处理方法等经验。显然，对制造过程实施连续监控中，当过程中再次出现这类已经出现过得、信号特征已知的异常信号（可称为"预期信号"），如果有效利用以往的积累下来的过程故障知识和处理经验，就可以大大提高对于过程异常的诊断速度和处理能力，但传统控制图显然无法做到这一点。Fisher[1]提出的 Cuscore 统计

---

① 金项目：国家自然科学基金资助项目（70672088）
作者简介：程志强（1968— ），男，河南安阳市人，南京理工大学博士生，从事质量工程与技术领域的研究；马义中（1964— ），男，河南泌阳人，南京理工大学教授、博士生导师，从事质量管理与工程领域的研究。

量在统计学中存在多年,以往 Box[2,3,4],Tsung F[5],Luceno A[6],SHU L[7]以及 Shao YE[8],Nembhard[9,10],J Ramirez[11]等人对于 Cuscore 统计量在质量管理中的应用研究多集中在过程监控或者 EPC/SPC 结合的问题上。重新考察 Cuscore 统计量在过程改进中的作用,将其整合到诸如六西格玛管理等管理模式中,以便通过 Cuscore 图来有效利用企业长期积累下来的过程异常知识和处理经验,提高过程质量诊断的效率,成为本研究的内容。

下面部分的内容安排顺序如下:第 2 部分,简述 Cuscore 检验统计量原理及其特点;第 3 部分,新提出更具一般性的非线性二次异常预期信号 $f(t) = x^2$,通过仿真,评价 Cuscore 图对其的检验性能;第 4 部分,针对提高改进效率的问题,提出在六西格玛管理 DMAIC(Define,Measurement,Analysis,Improvement,Control)模式中同时使用两张控制图进行质量监控的管理模式,并给出该模式的实现流程;第 5 部分,给出结语。

2　Cuscore 统计图

2.1　Cuscore 统计量

Box 等人认为,如果过程可以使用如下模型进行描述,
$$a_i = a_i(Y_i, X_i, \gamma), i = 1, 2, \cdots, t \tag{1}$$
其中,Y_i 是过程质量特性的观测值,X_i 是可控输入变量,γ 是待检测的预期异常信号的未知参数(如,阶跃信号的幅度;γ_0 为 γ 的初始值),a_i 是白噪声信号,$a_i \sim N(0, \sigma_a^2)$,方差 σ_a^2 已知但不依赖于 γ。则噪声样本信号的对数似然函数为:
$$L = -\frac{1}{2\sigma^2} \sum_{i=1}^{t} a_i^2 + c \tag{2}$$

对上式在 $\gamma = \gamma_0$ 微分并且令 $d_i = -\left.\frac{\partial a_i}{\partial \gamma}\right|_{\gamma=\gamma_0}$,则有:
$$\left.\frac{\partial L}{\partial \gamma}\right|_{\gamma=\gamma_0} = \frac{1}{\sigma^2} \sum_{i=1}^{t} a_{i0} d_i \tag{3}$$

Fisher 将累计和(Cuscore)统计量定义为：

$$Q_t = \sum_{i=1}^{t} a_{i0} d_i \tag{4}$$

其中，a_{i0} 为当 $\gamma = \gamma_0$ 时 a_i 的值(即,噪声的初始值)，通常称 d_i 为检验信号(detector)。

当 Cuscore 统计量用于监控过程信号时，使用下面公式监控 Cuscore 统计量的正向和负向变化，

$$\begin{cases} Q_t^+ = \max(0, Q_{t-1}^+ + a_{i0} d_t) \\ Q_t^- = \min(0, Q_{t-1}^- + a_{i0} d_t) \end{cases} \tag{5}$$

其中，$Q_0^+ = Q_0^- = 0$。Cuscore 控制图的工作原理是,如果 Q_t^+ 或 Q_t^- 超出了一定的限度,就认为过程出现了失控。

2.2 过程模型研究

Cuscore 统计量的具体形式是与过程的结构信息相关的。本文使用开环时间序列模型进行研究,不考虑反馈调节问题,其模型分为动态过程和扰动噪声过程两个部分,见图 1,这类似于生产过程中的单个工序情况。动态过程传递函数为 $G(B)$,其输入为 u_t;当不存在扰动 n_t 时动态系统的稳态输出目标值设定为 T;扰动 n_t 由预期信号 $f(t)$ 和使用 $ARMA(p,q)$ 表示的有色噪声 $\frac{\theta(B)}{\phi(B)} a_t$ 组成,即 $n_t = \frac{\theta(B)}{\phi(B)} a_t + \gamma f(t)$。其中,$B$ 是后向移位算子,满足 $B^K X_t = X_{t-K}$,$\phi(B)$ 和 $\theta(B)$ 分别表示为 $\phi(B) = 1 - \phi_1 B - \phi_2 B^2 - \cdots - \phi_P B^P$ 和 $\theta(B) = 1 - \theta_1 B - \theta_2 B^2 - \cdots - \theta_q B^q$,$a_t$ 的意义同上。

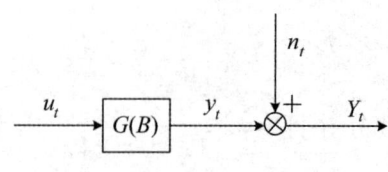

图 1 开环时间序列模型

显然,当动态过程达到稳态而预期信号 $f(t)$ 未出现时,

$$Y_t = T + \frac{\theta(B)}{\phi(B)} a_{t0} \tag{6}$$

当动态过程达到稳态且预期信号 $f(t)$ 出现时,

$$Y_t = T + \frac{\theta(B)}{\phi(B)} a_t + \gamma f(t) \tag{7}$$

假定 $\theta(B)/\phi(B)$ 可逆,对于式(7)求检验信号 d_t 可得,

$$d_i = -\frac{\partial a_i}{\partial \gamma}\bigg|_{\gamma=\gamma_0} = f(t) \frac{\phi(B)}{\theta(B)} \tag{8}$$

将式(6)、(8)代入式(4)可得上述模型的 Cuscore 统计量,

$$Q_t = \sum_{i=1}^{t} f(t) a_{t0} \frac{\phi(B)}{\theta(B)} \tag{9}$$

根据式(9)和(5),可以对图1中的模型绘制 Cuscore 控制图。

3 性能评价及仿真

评价 Cuscore 图的性能,可以使用识别率和失控平均链长 (Average Run Length,ARL) 两个指标,分别评价其对于预期异常信号的总体检测能力和检测速度。文献[15]使用 Cuscore 图对阶跃(Step)、斜坡(Ramp)等典型预期信号的检测能力进行了研究评价,而实际工业系统中的预期异常信号更多具有非线性特征,因此本文提出使用更具一般性的非线性二次预期信号 $f(t)$,考察 Cuscore 图对其的检测性能,

$$f(t) = \begin{cases} (t-t_0)^2, t > t_0 \\ 0, t \leqslant t_0 \end{cases} \tag{10}$$

设 γ 表示可以有不同取值的参数,显然,表达式 $f(t)$ 只是表示了预期信号的类型,$\gamma f(t)$ 则包含了预期信号的幅度变化快慢的信息。由于工业实际噪声动态过程多由一阶或二阶子系统组成[14],故本研究对图1中 $ARMA(p,q)$ 噪声过程模型选取 $p=q=1$。不失一般性,下面评价 Cuscore 图的检测性能时,假定预期信号 $f_i(t)$ 的出现时刻 $t_0 = 1$,是已知的;对于预期信号 $f_i(t)$ 的出现时刻未知的情况,解决方案参见[7,9]。

3.1 控制图绘制

使用式(9)和(5)绘制 Cuscore 图，控制限可以根据"控制图的受控 ARL 为 370"的概念，通过如下步骤的仿真方法确定，

(1) 生成 1000 个点的白噪声信号(即 a_{t0})。

(2) 使用公式(9)计算 Cuscore 统计量。

(3) 使用公式(5)得到 1000 个点的 Cuscore 图。

(4) 重复步骤(1)到(3)一万次，选取使得这一万张控制图的受控平均链长 ARL 为 370 的控制限作为 Cuscore 图的控制线。

根据上述原理，对于上述表达式(10)中的预期信号 $f(t)$ 通过蒙特卡罗(Monte Carlo)方法使用 MATLAB 软件进行仿真试验，绘制每张控制图包含 30 个点的 Cuscore 图和休哈特图，如图 2。

由图 2(C)看出，Cuscore 图在 $t=6$ 时就发出了异常报警信号；如果使用休哈特图对过程观测数据进行异常检测，依据 3sigma 控制限 CL=8.5，直到 $t=28$ 时才发出报警信号。

3.2 识别率及识别速度研究

本文在参数 ϕ_1 和 θ_1 的不同范围取值下研究 Cuscore 图对于式(10)表示的非线性预期异常信号的检测率和识别速度(失控 ARL)，并同休哈特图的性能表现相比较，1000 次仿真试验得到的平均结果见表 1 和表 2，其中括号中的数据为休哈特图的性能数据。

由试验结果可知，对于本文提出的二次非线性预期异常信号 $f(t)$，Cuscore 图可以实现 100% 检测，其检验能力不受参数 λ 的影响；而传统休哈特图的检测能力受参数 λ 及有色噪声模型参数 ϕ 和 θ 的影响较大，其对于预期信号幅度变化较小(λ 较小)的异常情况的检测率明显低于 Cuscore 图。检测速度方面，由失控 ARL 可知，Cuscore 图对于预期异常信号 $f(t)$ 的检测速度明显高于传统休哈特图。

附录 以第一作者身份发表的主要学术论文

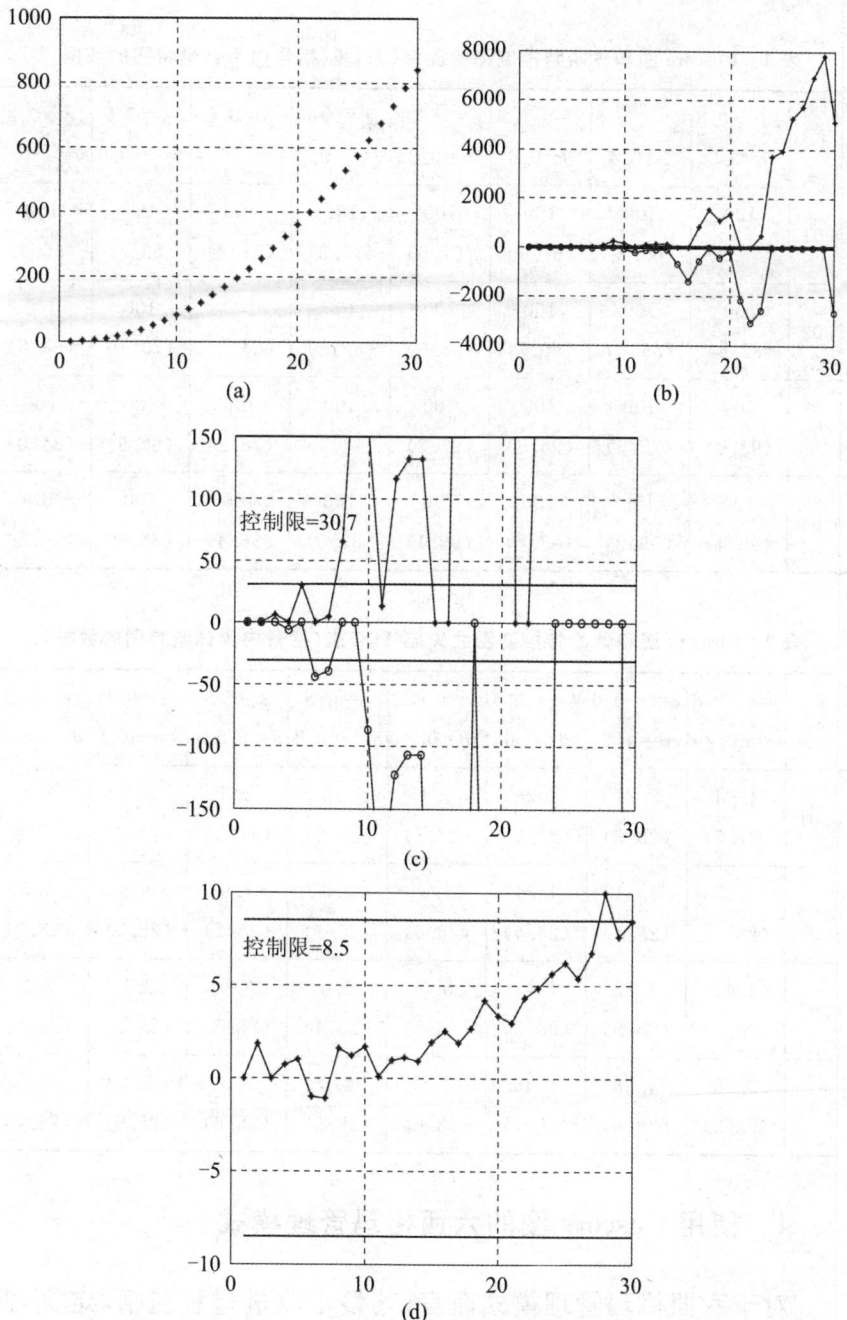

图 2 预期异常信号及其 Cuscore 图、休哈特图
(a)预期异常信号;(b)Q^+,Q^-统计图;
(c)带控制限的(Q^+,Q^-)双侧 Cuscore 图;(d)休哈特控制图

表1 Cuscore图和休哈特控制图检测率(%)表(括号内为休哈特图的数据)

λ	$\phi=-0.8$ $\theta=0.2$	$\phi=-0.6$ $\theta=0.4$	$\phi=-0.4$ $\theta=0.6$	$\phi=-0.2$ $\theta=0.8$	$\phi=0.8$ $\theta=-0.2$	$\phi=0.6$ $\theta=-0.4$	$\phi=0.4$ $\theta=-0.6$	$\phi=0.2$ $\theta=-0.8$
0.01	100 (53.2)	100 (60.8)	100 (68.1)	100 (67.6)	100 (47.3)	100 (51.8)	100 53.4	100 55.9
0.02	100 (82.4)	100 (86.3)	100 (90.7)	100 (88.0)	100 (62.3)	100 (64.2)	100 (70.10)	100 (68.8)
0.03	100 (93.2)	100 (95.9)	100 (96.8)	100 (97.3)	100 (71.4)	100 (78.2)	100 (80.6)	100 (84.0)
0.04	100 (98.4)	100 (99.2)	100 (97.8)	100 (99.1)	100 (82.0)	100 (86.4)	100 (86.10)	100 (89.3)

表2 Cuscore图和休哈特控制图之失控ARL表(括号内为休哈特图的数据)

λ	$\phi=-0.8$ $\theta=0.2$	$\phi=-0.6$ $\theta=0.4$	$\phi=-0.4$ $\theta=0.6$	$\phi=-0.2$ $\theta=0.8$	$\phi=0.8$ $\theta=-0.2$	$\phi=0.6$ $\theta=-0.4$	$\phi=0.4$ $\theta=-0.6$	$\phi=0.2$ $\theta=-0.8$
0.01	4.04 (27.7)	3.77 (28.8)	4.35 (27.8)	5.4 (27.8)	2.9 (26.5)	2.8 (27.4)	3.0 (27.7)	3.2 (27.8)
0.02	4.25 (28.2)	4.11 (28.3)	4.31 (28.4)	5.6 (28.4)	2.9 (28.1)	2.9 (28.2)	3.2 (28.3)	3.5 (28.3)
0.03	4.47 (28.5)	4.22 (28.5)	4.41 (28.5)	5.0 (28.5)	3.0 (28.4)	3.0 (28.5)	3.1 (28.5)	3.2 (28.5)
0.04	3.96 (28.5)	4.38 (28.6)	4.46 (28.7)	4.5 (28.5)	2.8 (28.5)	3.1 (28.6)	3.1 (28.6)	3.5 (28.7)

4 使用Cuscore图的六西格玛管理模式

对于六西格玛管理模式而言,其整个改进过程包括"定义,测量,分析,改进,控制"(DMAIC)五个阶段。本文提出将Cuscore控制图应用于六西格玛管理的新模式,见图3。该模式在分析阶段,对于过程异常进行分析总结,形成关于预期信号的表达式

$f_i(t)$以及相应的Cuscore算法知识库;在改进阶段,寻找由于异常信号$f_i(t)$造成的质量问题的解决方案并建立数据库;在质量改进的控制阶段,针对预期信号使用Cuscore图,一旦预期信号出现,就可以通过Cuscore图利用在前面专家数据库中积累的经验知识实现预期异常的迅速诊断和排除;同时,仍然使用传统SPC控制图对于未知原因和类型的质量异常进行监测并进行总结,逐渐归纳形成新的预期信号和积累处理经验并建立相应质量管理专家数据库。两类控制图同时使用,使得DMAIC模式可以有效利用积累下来的质量异常知识和处理经验,提高质量检测和故障处理水平。

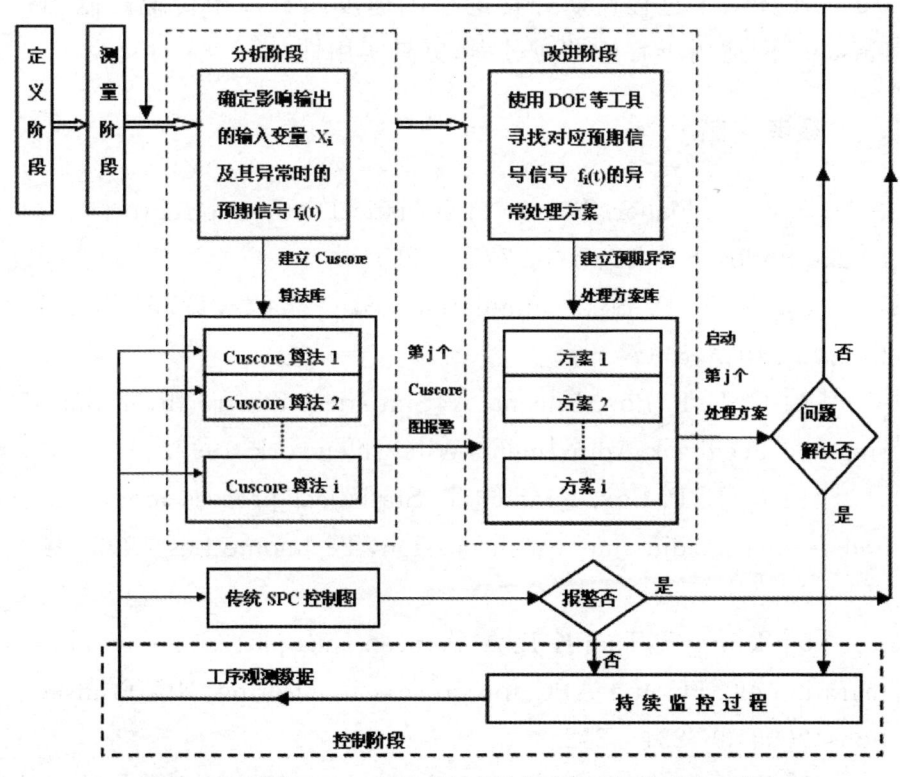

图3 同时使用Cuscore图和休哈特控制图的六西格玛管理模式

另外,传统控制图(休图,Cusum图,EWMA图)建立在过程观测数据的残差满足正态性的假设之上,实际的制造过程多数并不能真正满足;而 Cuscore 图并无这一要求,因而用于制造过程质量诊断更具普适性。

5 结论

研究及方针结果表明,同时使用两类控制图(Cuscore 图和传统休哈特控制图)对过程进行监控可以实现对于过程经验的有效利用,有效整合过程改进的模式,提高过程的诊断及处理效率;Cuscore 图对于本文提出的二次非线性预期信号的检测率明显高于传统休哈特控制图,并且不受该预期信号的幅度大小的影响,因此更适合于对于微小的过程预期异常信号进行检测;同时,Cuscore 图对于过程预期异常的检测速度明显高于传统控制图;Cuscore 图适用于有色噪声过程,更具实用性。

参考文献

[1] R. A. Fisher. Theory of statistical estimation, Proc. Cambridge Philos. Soc. 22,700－725(1925).

[2] G. E. P. Box. Cumulative score charts, Qual. Reliab. Eng. Int. 1992,8,17－27.

[3] G. E. P. Box. Luceno A, Statistical Control by Monitoring and Feedback Adjustment, Wiley:Newyork,1997.

[4] G. E. P. Box. Kramer T, Statistical process monitoring and feedback adjustment-a discussion, Technometrics 1992,34:251－285.

[5] Tsung F,Tsui K-L. A. A mean shift pattern study on integration of SPC and APC for process monitoring, IIE Transactions,2003,35:231－242.

[6] Luceno A. Cuscore charts to detect level shifts in autocorelated noise, Quality Technology and Quantitative Management,2004,1(1),27－45.

[7] Shu L,Apley,TSung F. Autocorrelated process monito-

ring using triggered cuscore charts, Qual. Reliab. Engng. Int. 2002,18:411—421.

[8] YE Shao. Integrated application of the cumulative score control chart and engineering process control, Statistica Sinica 1998,8:239—252.

[9] Herriet Black Nembhard, Pannapa Changpetch. Directed Monitoring using Cuscore Charts for seasonal Times Series, Qual. Reliab. Engng. Int. 2007,23:219—232.

[10] H. B. Nembhard, R. valverde-Ventura. Cuscore statistic to monitor a non-stationary system, Qual. Reliab. Engng. Int. 2007;23:303—325.

[11] J Ramirez. Monitoring clean room air using cuscore charts, Qual. Reliab. Engng. Int. 1992,14:281—289.

[12] Luceno A. Average run length and run length probability distributions for cuscore charts to control normal mean, Comput. Stat. Data Anal,1999,32:177—195.

[13] ZHAO NING HAN DONG. Chinese Journal of Applied Probability and Statistics,2005,Vol. 21 No. 4 NOV.

[14] H. B. Nembhard. Simulation using the state-space representation of noisy dynamic systems to determine effective integrated process control designs, IIE Transactions, 1998, 30: 247—256.

[15] Herriet Black Nembhard, Shuohui Chen. Cuscore Control Charts for Generalized Feedback-control systems, Qual. Reliab. Engng. Int. 2007;23:483—502.

Control Chart Pattern Recognition Based on PSO and LS-SVM[①]

Cheng Zhi-Qiang, Ma Yi-Zhong[②]

Department of Management Science and Engineering, Nanjing University of Science and Technology, Nanjing Jiangsu, 210094, P. R. China.

czqme@163.com

yzma-2004@163.com

Abstract: For the sake of reducing the cost of process monitoring, least squares support vector machines (LS-SVM) technique is introduced in this research to recognize the six typical patterns of control chart. The parameters of the LS-SVM pattern recognizer are optimized by particle swarm optimization (PSO) algorithm so as to improve the performance of LS-SVM pattern recognizer. Simulation results show that the optimized LS-SVM pattern recognizer can effectively recognize control chart patterns with high performance. The techniques introduced in this research can meet the requirements of process fault diagnosis under the condition of small sample size.

Key words: pattern recognition, least squares support vector machines, particle swarm optimization, control chart

---

[①] Supported by National Natural Science Foundation of China, Grant No. 70672088

[②] Corresponding author: yzma-2004@163.com, czqme@163.com

1 Introduction

By means of control chart pattern recognition, the normal/abnormal process status can be diagnosed. Pattern recognition technique is based on the data mining of process data. Guy [1,2] and Sukthomya, et al [3] conducted researches on this aspect using artificial neural network (ANN, such as BP, ART and SOM, etc.) and obtained some good results. But an obvious disadvantage of the ANN methods is that the methods require too many process data for training and testing the neural network. While manufacturing enterprises are unwilling to afford excessive testing cost anyway, how to use as less product samples as possible is the key consideration for the quality practitioners to select process fault diagnosis technique.

Support Vector Machine (SVM) [4] methodology trains the error of samples on the principle of "Structural Risk Minimization". The model structure of SVM depends only on the support vectors, and the training result of SVM has strong generalization ability. The above characteristics of SVM equip SVM recognizer with the ability of effectively recognizing the control chart patterns under the condition of small sample size. This trait of SVM technique is meaningful for the manufacturing process to reduce the cost of product test. Least squares support vector machines (LS-SVM) technique [5,6], which was proposed by Suykens, can transform quadratic programming problem of standard SVM into a problem of solvable linear equation. Thus, the computational speed of SVM algorithm can be improved effectively. However, LS-SVM technique has a problem of parameters selection when it is applied to control chart pattern recognition. The parameters of LS-SVM have significant effect on the performance of LS-SVM pattern recognizer, but the problem has not been an accepted way to obtain the optimal solution at present. Conventional solutions to the problem include cross-validation [7,8], leave-one-out [9], and the method of trial-and-error, etc. The aforementioned methods have more computation quantity and cannot get desired effect in most cases. Considering the importance of the LS-SVM parameters, this paper introduces "Particle Swarm Optimization" (PSO) [10,11,12] algorithm to optimize the parameters of LS-SVM recognizer. PSO algorithm has faster convergent speed compared with other evolutionary optimization algorithms. In the end, the six typical control chart patterns are recognized by the optimized LS-SVM pattern recognizer. Results of simulation experiment indicate that the optimized LS-SVM pattern recognizer has good pattern recognition performance. The simulation results are also compared with the results of other similar researches [1,13] in which BP and PCA-SVM methods were used respectively.

2 Principle of least squares support vector machines

Let x_i be a set of data points and y_i indicate the corresponding classes, where $x_i \in R^n$, $y_i \in R$, and $i = 1, \cdots, N$. LS-SVM technique utilizes $\sum_i^N \xi_i^2$ to substitute $\sum_i^N \xi_i$ of the standard SVM, and the optimization problem is formulated as following,

$$\begin{cases} \min \quad J(\omega,\xi) = \frac{1}{2}\omega^T\omega + \frac{1}{2}\gamma\sum_{i=1}^N \xi_i^2 \\ \text{s. t.} \quad y_i = \omega^T\phi(x_i) + b + \xi_i \end{cases} \quad (1)$$

where $\phi(\cdot)$ is a nonlinear mapping function. Using Lagrange multipliers technique and KKT complementary condition, we can transform the above optimization problem into the problem of solving the following linear equation,

$$\begin{bmatrix} \Omega + \gamma^{-1}I & l_N \\ l_N^T & 0 \end{bmatrix} \begin{bmatrix} \alpha \\ b \end{bmatrix} = \begin{bmatrix} y \\ 0 \end{bmatrix} \quad (2)$$

where $l_N = [1,1,\cdots,1]^T$, $\alpha = [\alpha_1,\alpha_2,\cdots,\alpha_N]^T$ and $y = [y_1, y_2,\cdots,y_N]^T$. Ω is a square matrix with entries $\Omega_{ij} = \psi(x_i,y_j) = \phi(x_i)^T \phi(x_i)$, $i,j = 1,\cdots,N$. $\psi(\cdot)$ is a kernel function which satisfies the Mercer's theorem. α and b can be obtained by solving equation (2), and the linear regression model can be represented by the following formula:

$$y(x) = \sum_{i=1}^{N} \alpha_i \psi(x,x_i) + b \quad (3)$$

In this research, Gaussian kernel function is chosen as ψ, so the expression of ψ is expressed as following,

$$\psi(x,y) = \exp(\frac{-\|x-y\|^2}{2\sigma^2}) \quad (4)$$

Finally, the resulting LS-SVM model for function estimation becomes [14],

$$f(x) = \sum_{i=1}^{N} y_i \alpha_i \psi(x,x_i) \quad (5)$$

And the final decision function becomes $\text{sgn}[f(x)]$, where $\text{sgn}(x) = \begin{cases} 1, x > 0 \\ -1, x < 0 \end{cases}$.

3 The optimization of LS-SVM parameters through PSO

When LS-SVM technique is used for control chart pattern recognition, we need to optimize the parameter γ in formula (1) and the parameter σ in formula (2). The two parameters are the key factors that influence the performance of LS-SVM recognizer. This paper proposes to utilize PSO algorithm to optimize the two parameters of LS-SVM recognizer. PSO algorithm can transfer the information of the current global optimum solution to all the particle individuals, so the particle swarm can converge at the optimum solution more rapidly. PSO algorithm has the advantage of evolution policy in most cases.

3.1 PSO theory

PSO algorithm was discovered through simulation of social behavior of a school of flying birds [10]. As for PSO, a feasible solution corresponds to a position of a "bird" in the search space of the optimization problem. The bird is the so-called "particle". When the particle swarm is initialized, a number of initial feasible solutions of the optimization problem are generated. Every individual particle has a fitness value which is determined by optimizing the objective function. Meanwhile, the velocity of every particle individual can determine the fly direction and the fly distance for the next generation of particles. In the process of searching global optimal solution, every particle tails after both the optimum solution (p_{iK}) of its own and the optimal solution (p_{gK}) of the particle population. Every particle updates its velocity (v_K) and position (x_K) through the following two formulae:

$$v_{K+1} = wv_K + c_1(p_{iK} - x_K) + c_2(p_{gK} - x_K) \qquad (6)$$

$$x_{K+1} = x_K + v_{K+1} \qquad (7)$$

As for the optimization problem of LS-SVM recognizer, x_K is the current value of parameters $\{\gamma, \sigma\}$ in the course of iterative calculations.

3.2 Parameters setting of PSO

1) Parameters Setting of PSO.

In formula (6), c_1 and c_2 are two accelerative factors in the dynamic range of (0,2). The maximum velocity of every particle is set as $\nu_{max} = 2$, $\nu_K \in [-\nu_{max}, \nu_{max}]$. Parameter w is the inertia weight. In order to reduce the number of iterations, we can modify w through the following formula (8) [11,12] in the course of iterative calculations.

$$w = w^0 - (w^0 - w^\infty) \times K/(MS) \qquad (8)$$

where w^0 is the initial inertia weight, w^∞ is the final inertia weight, M is the maximum iteration number, K is the current iteration number, S is a coefficient. In this paper, let $w^0 = 0.95$, $w^\infty = 0.4$, $M = 50$, $S = 0.7$.

2) Design of particle fitness value.

The two parameters γ and σ of LS-SVM pattern recognizer should be optimized in total in the course of evolutionary iteration. One of the emphases of this research is the performance of the LS-SVM recognizer, so this research chooses the pattern recognition rate as the fitness value of the particle population.

3) Initialization of particle population.

The Initialization of particle population comprises two aspects, position and velocity. Population size is 10 in this paper. When the initial particle population is generated, the positions and the velocities of particle individuals are randomly generated in the range of (0,20) according to experience.

4 Data sets of control charts and pattern recognition project design

There are six kinds of control chart patterns [1,13] need to be recognized in this research. They are illustrated in Figure 1.

The six pattern datasets can be represented by the following equations,

For pattern data set $x(t)$,

· 199 ·

$$x(t) = \mu + d(t) + r(t) \tag{9}$$

where t denotes time, $x(t)$ is the sample value at time t, μ is the mean value of the process variable, $r(t)$ is the common cause variation of the process at time t, $r(t) \sim N(\mu, \sigma^2)$, $d(t)$ is the special cause disturbance at time t. Without loss of generalization, let $\mu = 0$, $\sigma = 1$ in this paper. The type of control chart pattern is decided by $d(t)$.

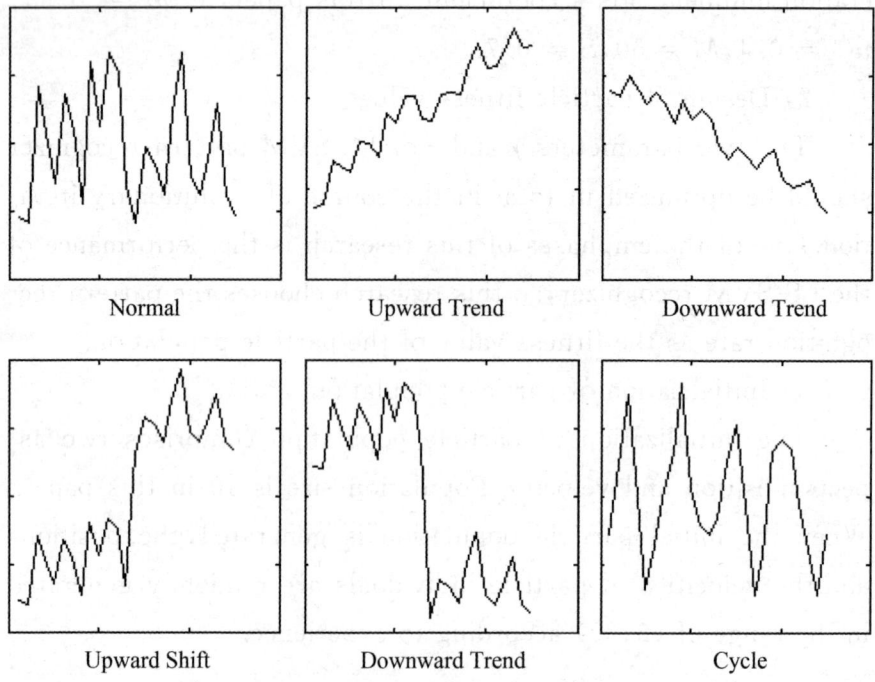

Fig 1　Six Kinds of Control Chart Patterns

1) For normal pattern,
$$d(t) = 0 \tag{10}$$

2) For upward/downward trend pattern,
$$d(t) = kt \tag{11}$$

where k is the trend slope in terms of σ ($0.15\sigma \leqslant k \leqslant 0.3\sigma$, $-0.3\sigma \leqslant |k| \leqslant -0.15\sigma$).

3) For upward/downward shift pattern,

$$d(t) = 1(t-t_0)s \qquad (12)$$

where t_0 is the parameter to determine the shift position of $d(t)$ ($1(t-t_0) = 1$ when $t \geqslant t_0$, $1(t-t_0) = 0$ when $t < t_0$). s is the shift magnitude in terms of σ ($\sigma \leqslant s \leqslant 3\sigma, -3\sigma \leqslant s \leqslant -\sigma$).

4) For cyclic pattern,

$$d(t) = A\sin(2\pi t/T) \qquad (13)$$

where A is the cycle amplitude in terms of σ, $1.5\sigma \leqslant A \leqslant 3\sigma$. T is the cycle period, $2 \leqslant T \leqslant 8$, let $T = 7$ in this paper.

4.1 Dataset preparation

In order to decrease the effect of the data noise on the performance of LS-SVM recognizer, we should prepare the simulation/process data before they are input into LS-SVM recognizer. The data preparation scheme can be summarized as follows,

$$z(t) = (x(t) - E(\mu))/E(\sigma) \qquad (14)$$

$$\begin{cases} y(t) = 25, z(t) \geqslant 4.9 \\ y(t) = -25 + N, -4.9 + 0.2(N-1) \leqslant z(t) < -4.9 + 0.2N \\ y(t) = -25, z(t) < -4.9 \end{cases}$$
$$(15)$$

where N is the parameter to determine the coding rate, $N = 1, 2, 3, \cdots, 49$. $E(\mu)$ and $E(\sigma)$ are the estimated values of μ and σ respectively. As for the simulation data of this paper, we set $E(\mu) = 0, E(\sigma) = 1$. $y(t)$ is the coded value of $x(t)$.

4.2 Pattern recognition project design

LS-SVM technique can only construct pattern recognizer used for two-class datasets. As for the multi-class classification problem, the "one against one" [15] classification project is used for constructing a multi-class pattern recognizer. Let K be an integer greater than 2, this project firstly constructs C_K^2 two-class

LS-SVM recognizers where each one is trained on dataset from two different classes, where $C_K^2 = K(K-1)/2$. Thus, according to the LS-SVM technique in formula (1), we can obtain C_K^2 two-class recognizer decision function. For training data from the mth and the nth classes, the recognizer decision function for class m and class n is defined by $(\omega_{mn}^T \phi(x) + b_{mn})$. For a testing sample x_t of control chart, the following voting strategy is used: if $sgn(\omega_{mn}^T \phi(x_t) + b_{mn})$ shows x_t is in the mth class, then the vote for the mth class is added by one. Otherwise, the nth is increased by one. Finally, x_t is predicted to be in the class with the largest vote. Namely, let $f_{mn}(x) = \omega_{mn}^T \phi(x) + b_{mn}$ and $f_n(x_t) = \sum_{m \neq n, m=1}^{K} sgn(f_{mn}(x_t))$, then we classify x_t into the following class

$$\operatorname*{argmax}_{n=1,\cdots,K} f_n(x_t) \tag{16}$$

The one-against-one project is an extension of the two-class LS-SVM technique. The voting approach described above is the so-called 'Max Wins' strategy.

5 Simulation experiment results

5.1 Simulation experiment procedure

Simulation experiment of control chart pattern recognition was performed in this research with LS-SVM technique and PSO algorithm. The simulation procedure employs the following five steps.

Step 1: Randomly generate 10 $\{\gamma, \sigma\}$ particles as the initial position of the particle population, and initiate the velocity of particle population in the same way. Call "Control Chart Pattern Recognition Subprogram" shown in Figure 2 to compute the fitness values of the particle individuals, as well as the global optimum position and global optimum fitness value for the initial particle population.

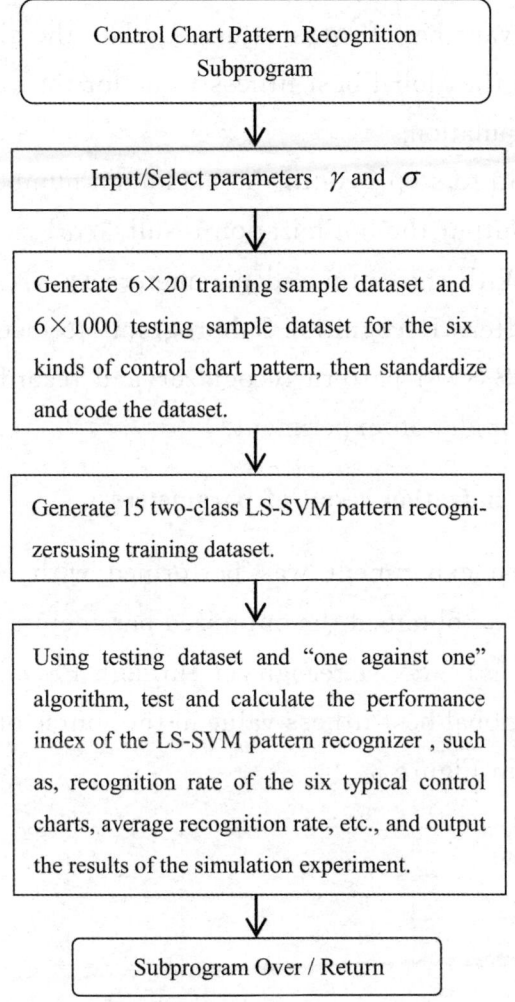

Fig 2　Flow Chart of "Control Charts Pattern Recognition Subprogram"

Step 2: Use formula (8),(6) and (7) to update the velocities and positions of particle individuals so as to form the new generation of particle swarm $\{\gamma,\sigma\}$. Call "Control Chart Pattern Recognition Subprogram" to compute the fitness values of particle individuals for the new generation of particle population.

Step 3: According to the new fitness values of particle individuals, update the best positions p_{iK} of new particle individuals and their relevant best fitness values. Update the global best position p_{gK} and the global best fitness value for the new generation of particle population.

Step 4: Go to step 2, until the maximum number of iteration is achieved. Output the optimization result $\{\gamma,\sigma\}_{best}$.

Step 5: With the optimized parameters $\{\gamma,\sigma\}_{best}$, call "Control Chart Pattern Recognition Subprogram" to test the performance of the LS-SVM pattern recognizer, and record the final results of the simulation experiment.

5.2 Optimization result of parameters

Simulation experiment was performed with MATLAB7.4 software, and we obtained the optimized parameters $\gamma = 19.6154$, $\sigma = 8.4538$ for LS-SVM recognizer through PSO. The variation curve of the global best fitness value in the course of 50 iterations is illustrated in Figure 3.

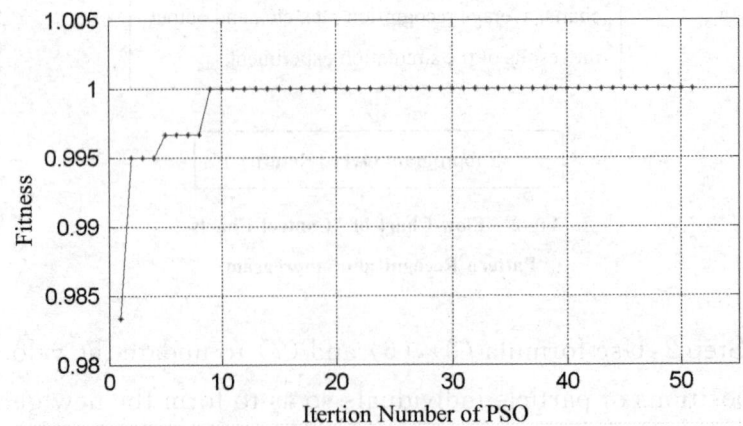

Fig 3 Variation Curve of the Global Best Fitness Value

In Figure 3, the first point (at $t=1$) represents the initial value of the global best fitness, the global best fitness value tends to be stable (at $t=9$) after only 8 iterations.

5.3 Experiment results of control chart patterns recognition

When the simulation experiment was performed, 20 training samples of each kind of control chart were used for training the LS-SVM recognizer, 1000 testing samples of each kind of control chart were used for testing the performance of the LS-SVM recognizer. The experiment results of the control chart pattern recognition are listed in Table 1.

Tab 1 Results of the control chart pattern recognition

单位:%

Pattern type	Pattern recognition rate of six kinds of control chart						Average recognition rate
	Normal	Upward Trend	Downward Trend	Upward Shift	Downward Shift	Cycle	
Results	97.8	99.4	99.4	99.6	99.8	99.8	99.3
Results of [13]	99.25	96.5	96.5	96.25	95.75	100	97.4
Results of [1]		86	89	90	89	94	89.6

From the experiment results in Table 1, it can be concluded that the optimized LS-SVM pattern recognizer can work with high performance according to pattern recognition rate. Comparing with the PCA-SVM technique [13], LS-SVM technique can avoid the process of principal component analysis for each training sample and obtain benefit from low computation cost. Comparing with the BP neural network technique [1], LS-SVM technique uses only $20 \times 6 = 120$ training samples, the amount of training samples is greatly less than that of neural network technique. LS-SVM technique also has an advantage over neural network technique in training speed.

6 Conclusion

The proposed LS-SVM technique can effectively recognize the patterns of the six typical control charts with high performance under the condition of small sample size. The performance of LS-SVM pattern recognizer can be improved by means of PSO algorithm. LS-SVM technique is based on the principle of structural risk minimization (SRM), so we do not have to consider the normality of the process data when LS-SVM technique is used for monitoring the process status. The characteristics of the LS-SVM technique are of great economic value in practical application. The results of this research are helpful for the enterprises to reduce the cost of process fault diagnosis.

References

[1] Ruey-Shiang Guh. Robustness of the neural network based control chart pattern recognition system to non-normality. International [J]. Journal of Quality & Reliability Management, 2002,19(1):97—112.

[2] Ruey-shiang Guh. A hybrid Learning-based Model for On-line Detection and Analysis of Control Chart Patterns [J]. Computers & Industrial Engineering,2005,49:35—62.

[3] Wimalin Sukthomya, James Tannock. The training of neural networks to model manufacturing processes [J]. Journal of Intelligence Manufacturing 2005,16:39—51.

[4] Nello Cristianini, John Shawe-Taylor. An Introduction to Support Vector Machines and Other Kernel-based Learning Methods [M]. Cambridge University Press,2005.

[5] Suykens Johan A K, Gestel Tony Van, Brabanter Jos

De. Least Squares Support Vector Machines [M]. Singapore: World Scientific Publishers, 2005.

[6] Hoegaerts L., Suykens, J. A. K. Vandewalle. A comparison of pruning algorithms for sparse least squares support vector machines [J]. Neural Information Processing. 11th International Conference, ICONIP 2004, 1247－1253.

[7] V. G. Tony, et al. Benchmarking Least Squares Support Vector Machine Classifiers [J]. Machine Learning, 2004, 54(1): 5－32.

[8] Senjian An, Wanquan Liu, Venkatesh, S. Fast cross-validation algorithms for least squares support vector machine and kernel ridge regression [J]. Pattern Recognition, 2007, 40(8): 2154－62.

[9] G. C. Cawley, N. L. C. Talbot. Fast exact leave-one-out cross-validation of sparse least-squares support vector machines [J]. Neural Networks, 2004, 17(10): 1467－75.

[10] Kennedy J, Eberhart R. Particle swarm optimization [A]. Proceeding IEEE international conference on neural networks [M], 1995, 4: 1942－1948.

[11] Shi Y, Eberhart R. A modified particle swarm optimizer [A]. IEEE World Congress on Computational Intelligence. 1998, p 69－73.

[12] Yannis Marinakis, Magdalene Marinaki, Georgios Dounias. Particle swarm optimization for pap-smear diagnosis [J]. Expert Systems with Applications, 2008, 35: 1645－1656.

[13] YANG Shi-yuan, WU De-hui, SU Hai-tao. Abnormal Pattern Recognition Method for Control Chart Based on Principal Component Analysis and Support Vector Machine [J]. Journal of System Simulation, 2006, 18(5): 1314－1318.

[14] C. L. Huang, C. J. Wang. A GA-based feature selection

and parameters optimization for support vector machines [J]. Expert Systems with Applications 2006,31:231—240.

[15] Bo Liu, Zhifeng Hao, Xiaowei Yang. Nesting algorithm for multi-classification problems [J]. Soft Computing, 2007, 11: 383—389.

◀ 附录　以第一作者身份发表的主要学术论文

Mean Shifts Identification Model in Bivariate Process Based on LS-SVM Pattern Recognizer

Cheng Zhi-Qiang[1], Ma Yi-Zhong[1], Bu Jing[2]

[1]*Department of Management Science and Engineering, Nanjing University of Science and Technology, Nanjing Jiangsu, 210094, P. R. China.*

czqme@163.com, yzma-2004@163.com

[2]*Automation Institute, Nanjing University of Science and Technology, Nanjing Jiangsu, 210094, P. R. China.*

bujing30@foxmail.com

Abstract: This study develops a least squares support vector machines (LS-SVM) based model for bivariate process to diagnose abnormal patterns of process mean vector, and to help identify abnormal variable(s) when Shewhart-type multivariate control charts based on Hotelling's T^2 are used. On the basis of studying and defining the normal/abnormal patterns of the bivariate process mean shifts, a LS-SVM pattern recognizer is constructed in this model to identify the abnormal variable(s). The model in this study can be a strong supplement of the Shewhart-type multivariate control charts. Furthermore, the LS-SVM techniques introduced in this research can meet the requirements of process abnormalities diagnosis and causes identification under the condition of small sample size. An industrial case application of the proposed model is provided. The performance of the proposed model was evaluated by computing its classification accuracy of the LS-SVM pattern recognizer. Results from simulation case studies indicate that the proposed model is a successful method in identifying the abnormal variable(s) of process mean shifts. The results demonstrate that the proposed method provides an excellent performance of abnormal pattern recognition. Although the proposed model used for identifying the abnormal variable(s) of bivariate process mean shifts is a particular application, the proposed model and methodology here can be potentially applied to multivariate SPC in general.

Key words: multivariate statistical process control, least squares support vector machines, pattern recognition, quality diagnosis, bivariate process

1 Introduction

In many industries, complex products manufacturing in particular, statistical process control (SPC) is a widely used tool of quality diagnosis, which is applied to monitor process abnormalities and minimize process variations. According to Shewhart's SPC theory, there are two kinds of process variations, common cause variations and special cause variations. Common cause variations are considered to be induced by the inherent nature of normal process. Special cause variations are defined as abnormal variations of process, which are induced by assignable causes. Traditional univariate SPC Control charts are the most widely used tools to reveal abnormal variations of monitored process. Abnormal variations should be identified and signaled as soon as possible to the effect that the quality practitioners can eliminate them in time and bring the abnormal process back to the normal state.

In many cases, the manufacturing process of complex products may have more than two correlated quality characteristics and a suitable method is needed to monitor and identify all these characteristics simultaneously. For the purpose of monitoring the multivariate process, a natural solution is to maintain a univariate chart for each of the process characteristics separately. However, this method could result in higher fault abnormalities alarms when the process characteristics are highly correlated [1] (Loredo, 2002). This situation has brought about the extensive research performed in the field of multivariate quality control since the 1940s, when Hotelling introduced that the quality of a product may depend on several correlated characteristics in a multivariate manufacturing environment. From then on, many researchers have proposed several multivariate statistical control charts (MSPC), where each has some advantages as well as disadvantages. Traditional MSPC techniques include Hotelling T^2 control charts [2] (Aparisi and Haro, 2001), MEWMA control charts [3,4] (Lowry and Montgomery 1995; Mason et al. 1997) and CUSUM control charts [5] (Crosier, 1988). Those MSPC techniques can be used to monitor the process shifts of mean vector. In a manufacturing environment, Hotelling's T^2 statistic is the optimal test statistic for finding a general shift in the process mean vectors for multivariate process; it is not optimal for shifts that occur for some subset of variables. The advantage of MEWMA control charts over the Hotelling T^2 control charts is that their average run length (ARL) is smaller for small shifts in the process mean. The properties of MCUSUM control charts are quite similar to those of the MEWMA charts. In order to improve the performance of detecting small shifts in multivariate processes, Prabhu and Runger (1997) [6] gave some suggestions in the selection of the parameters of a MEWMA control chart. There exist some other techniques for controlling multivariate processes, such as multi-way PCA [7] (Louwerse and Smilde, 2000) and multi-way PLS [8] (Nomikos and MacGregor, 1995), the application of them is to reducing the dimensionality of the variable space for a multivariate process. For a review of multivariate control charts, readers are referred to the literature [9] (Bersimis, Psarakis, and Panaretos, 2007). Traditional MSPC techniques evaluate the normal or abnormal state of a process based upon an overall statistic, this leads to a persistent problem in multivariate control charts, namely the interpretation of a signal that often discourages practitioners in applying them [10] (Alt, 1985). When traditional MSPC charts alarm that a process is out of control, they do not provide information on which variable or group of variables is out of control, that is to say, they can not tell the quality practitioners the answer to the question "What component(s) of the mean vector are the cause(s) of the alarm signal?". For monitoring process mean shifts, some researchers tried to determine which of the process variables (or which subset of them) is responsible for the process abnormal signal by using the decomposition method [11,12] (Mason & Young, 1999; Runger, Alt and Montgomery, 1996). However, little work has been done on identifying the sources of process abnormality.

In the last decades, various artificial neural networks (ANN) techniques have been applied into the abnormalities detection of process. ANN techniques require no hypothesis on statistical distribution of the monitored process observations. This significant attribute makes ANN useful and effective tools that can be used to improve data analysis in process quality diagnosis applications. Especially in recent years, a lot of research has focused on the effective application of ANN to detect process abnormalities or to identify the source(s) of a process abnormality in multivariate processes. Ho and Chang (1999) [13] presented an ANN model for monitoring the process mean and variance shifts simultaneously and classifying the types of shifts. Noorossana (2003) [14] presented an ANN model to effectively detect and identify the source(s) of a process abnormality in auto-correlated processes. Niaki (2005) [15] successfully trained a multi-layer perceptron (MLP) network to detect shifts in mean when Hotelling T^2 control charts alarmed a abnormal process signal, and this network could also locate and identify the abnormal variable(s). Guh (2007)[16] and Yu(2009)[17] developed a back-propagation network model to identify and quantify the mean shifts in bivariate processes. The proposed model outperforms the traditional multivariate control charts in terms of ARL, and can accurately estimate the magnitude of the shift of each of the shifted variables. The major results of those cited ANN-based methods indicate that ANN models are useful and effective methods for abnormal variable(s) identification and cause-selection problems. But ANN models have common issues of selecting the structure of a neural network (for instance, the number of layers and the corresponding number of neurons in every layer) and the adjusting of the parameters of ANN algorithm (for instance, learning rate). Researchers usually have to solve those problems according to their personal experiences and can not achieve good effects. In addition, under the modern market environment of competition, customers have high and diversified demands for complex products manufacturing systems, modern manufacturing processes accordingly have the feature of small batch production [18] (Lin et al, 1997). For example, the famous flexible manufacturing systems (FMS) and just-in-time (JIT) production systems are characterized by small batch sizes. In such cases, it is very difficult to get mass test data from process with high testing cost. The objective conditions of today ask quality practitioners to identify and locate the causes of process abnormity under the condition of small sample size. But traditional MSPC methodology is based on the large sample statistical theory, the validity of the process diagnosis can only be guaranteed on the basis of large sample size. Moreover, traditional MSPC techniques demand that the process observations conform to normality assumption, and this is very difficult to fulfill strictly in many manufacturing processes. Since ANN techniques need large amounts of training sample data when they are used for multivariate process monitoring, ANN techniques can not be applied to diagnose the process abnormalities under the condition of small batch production. At the same time, almost all the manufacturing enterprises are unwilling to afford excessive cost for obtaining large amounts of test data anyway.

To solve the above-mentioned problems which exist in process diagnosis, this work proposes a bivariate process diagnosis model to locate and identify the abnormal variable(s) of the process mean shifts. Mean shifts are regarded as sudden shifts in mean vector that present an abrupt change in levels within a series of observations. In this bivariate process diagnosis model, the problem about abnormal variable(s) identification and location is converted into the problem about abnormal pattern recognition of process mean vector. Least square support vector machine (LS-SVM) technique is a kind of machine learning algorithm based on statistical learning theory, it is powerful in machine learning to solve pattern classification problems with limited number of samples. This model employs LS-SVM technique to construct pattern recognizer of bivariate processes mean vector, and the function of this model is to locate and identify the abnormal mean shift variable(s) of bivariate process under the circumstance of small sample size. Although this abnormal variable(s) identification model for bivariate processes is the particular application presented here, the proposed model and methodology can be potentially applied to multivariate SPC in general. As an application, we present a real world case to illustrate the usage of the proposed SVM-based pattern recognition model for identifying the abnormal variable(s) of bivariate process mean shifts. The performance of the model is evaluated by performance index of identification rates of abnormal mean shifts variable(s) in this case. The results indicate that the proposed model performs efficiently for a bivariate process to identify the mean shifts.

The rest of this paper is structured as follows. Following the introduction, Section 2 proposes a LS-SVM-based model for locating and identifying the abnormal mean shifts variable(s) in bivariate process. Section 3 briefly reviews the theory of Hotelling's T^2 control charts. Section 4 introduces LS-SVM technique used for recognizing the normal/abnormal patterns of bivariate process mean shifts. Section 5 describes the way to generate the training data and testing data for the diagnosis model, studies a real world case through numerical experiments and exposes the obtained results. Finally, Section 6 presents a concluding summary and suggests some directions for future research.

2 LS-SVM Based Model to Recognizing the Mean Shifts Patterns in Bivariate Process

In order to identify and locate the abnormal variable(s) for the bivariate process mean vector under the condition of small sample size, this research proposes to construct a bivariate process diagnosis model based on Hotelling T^2 and LS-SVM pattern recognizer, and the model structure is schematically shown in Figure 1. This model include two sections, the first section employ Hotelling T^2 statistic to detect the overall out-of-control signal for the bivariate process. When Hotelling T^2 statistic signal a alarm, the LS-SVM pattern recognizer in the second section may model the cause-selecting problem as a pattern classification problem, namely, the abnormal signal can be classified into several classes of patterns. As for the bivariate case with mean vector $\mu = (\mu_1, \mu_2)^T$ and covariance matrix $\sum$, there are three distinguished classes of abnormal patterns: (1) the first variable being out of control, (2) the second variable being out of control, and (3) the first and second variables are both out of control. It is obvious that the abnormal signal must belong to one of them. By this way, the LS-SVM pattern recognizer can supply the accurate information about the abnormal variable(s) for taking corrective action.

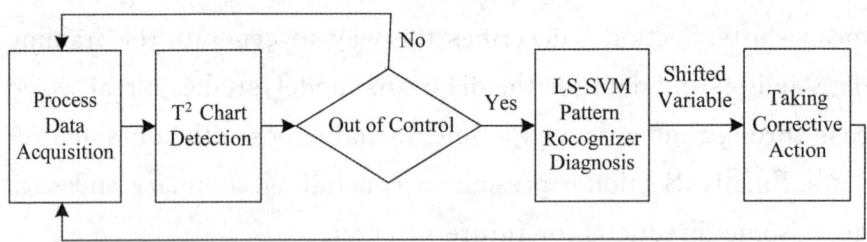

Fig 1 Process Diagnosis Model by T^2 Chart and LS-SVM Pattern Recognizer

This model suggests that, for a P-dimensional manufacturing process with p variables, the mean of every variable has two states: normal state, and abnormal state. Consequently, the P-dimensional process has 2^P possible normal/abnormal states. It is no doubt that only one of 2^P states is the normal state. When T^2 statistic has detected the abnormality in the process, there are ($2^P -$

1) different abnormal states altogethers. Please note that once the abnormal state occurred in the process mean, it belongs and only belongs to one of the (2^P-1) abnormal states. For the purpose of discriminating the (2^P-1) possible abnormal states from each other, the diagnosis model in this research defines the (2^P-1) abnormal states as (2^P-1) abnormal patterns which need to be recognized. When the T² statistic give alarm signal, the proposed model can employ the trained LS-SVM pattern recognizer to classify the abnormal patterns of the process mean vector and further more identify the abnormal component(s) of the mean vector. For example, if denoting the abnormal variable as 1 and the normal variable as 0, then the patterns of the mean vector for bivariate process can be indicated by (0,0), (0,1), (1,0) and (1,1) respectively, where (0,0) represents the normal pattern of the process mean vector, and (0,1), (1,0) and (1,1) represent the other three abnormal patterns of the process mean vector. When T² statistic signal an alarm, if the LS-SVM pattern recognizer classify the abnormal pattern into (0,1), it can be identified that only the second component of process mean vector is the cause of the process abnormality; else if LS-SVM pattern recognizer classify the abnormal pattern into (1,1), it can be identified that both the two components of process mean vector are the causes of the process abnormality.

3 Hotelling T² Control Chart

As for a P-dimensional process with samples $\{X_1, X_2, \cdots, X_n\}$, $X_i = (X_{i1}, X_{i2}, \cdots, X_{iP})^T$ represents the i-th P-dimensional vector of measurements made on a process at time period i when the process is in control, with X_{ij} the i-th individual observation of the j-th characteristic. Usually, $n > P$ is required in the research. Hotelling T^2 control charts assume that when the process is statistically in control, X_i are independent and follow a multivariate normal distribution with mean vector μ and covariance matrix $\sum$, that is, $X_i \sim N(\mu, \sum)$. For individual multivariate observations, when μ and $\sum$ are unknown in the practical application, these parameters can be estimated from n reference samples. The usual parameter estimators are the sample mean vector $\overline{X}$ and the sample covariance matrix S, where $\overline{X} = (\overline{X}_1, \overline{X}_2, \cdots, \overline{X}_P)^T$, $S = \frac{1}{n-1} \sum_{i=1}^{n} (X_i - \overline{X})(X_i - \overline{X})^T$, $\overline{X}_j = \frac{1}{n} \sum_{i=1}^{n} X_{ij}$. Then the T^2 statistic for X_i can be constructed in the following manner:

$$T_i^2 = (X_i - \overline{X})^T S^{-1} (X_i - \overline{X}) \qquad (1)$$

which, when multiplied by the constant $\frac{n(n-P)}{P(n^2-1)}$, follows a F distribution with P and $(n-P)$ degrees of freedom. Accordingly, In Hotelling T^2 chart, for each observation X_i, the T^2 statistic is calculated and checked by the upper control limit (UCL), that is given by the following formula at certain significance level α,

$$UCL_{T^2} = \frac{P(n^2-1)}{n(n-P)} F_\alpha(P, n-P) \qquad (2)$$

$$LCL_{T^2} = 0 \qquad (3)$$

In particular, if the process parameters (i.e., μ and $\sum$) are already known, then χ^2 is used instead of T^2 statistics, the UCL in this case is $\chi^2_{\alpha,p}$ [19]; Otherwise the T^2 statistics itself which follows the F distribution is used. In effect, T^2 represent the weighted distance of any point from the target (i.e., the process mean under stable conditions). The Hotelling T^2 chart can point out that the process moved to an out-of-control state when the T_i^2 value of observation X_i exceeds the control limit UCL_{T^2}. Please note that, like any other statistical quality control procedure, there are always two kinds of errors, namely type I (α) and type II (β).

4 LS-SVM Pattern Recognizer

Support vector machine (SVM) [20,21] (Vapnik 1995; Nello 2005) is a kind of machine learning method based on statistical learning theory, which mainly researches the learning of limited samples. As a new and promising classification and regression technique, the structure of SVM model is only determined by the so-called "support vectors", so SVM has strong ability of data classification under the condition of small sample size. Least squares support vector machines (LS-SVM) [22] (Suykens, 1999) is extended model of standard SVM. SVM are trained by solving a quadratic optimization problem, while LS-SVM is trained by solving a set of linear equations, so LS-SVM method has faster computational speed. In this research, the LS-SVM method is applied to construct the pattern recognizer for the bivariate mean vector. A simple description of LS-SVM method is provided below, for more details please refer to literature [23] (Suykens et al., 2002).

4.1 LS-SVM Principle

Let x_i be a set of data points and y_i indicate the corresponding classes, where $x_i \in R^n, y_i \in R$, and $i = 1, \cdots, N$. LS-SVM technique can formulate the optimization problem as following,

$$\begin{cases} \min \quad J(\omega, \xi) = \frac{1}{2}\omega^T\omega + \frac{1}{2}\gamma \sum_{i=1}^{N} \xi_i^2 \\ \text{s. t.} \quad y_i = \omega^T \phi(x_i) + b + \xi_i \end{cases} \quad (4)$$

where $\phi(\cdot)$ is a nonlinear mapping function. And the above optimization problem can be finally transformed into the problem of solving the following linear equation,

$$\begin{bmatrix} \Omega + \gamma^{-1}I & l_N \\ l_N^T & 0 \end{bmatrix} \begin{bmatrix} \alpha \\ b \end{bmatrix} = \begin{bmatrix} y \\ 0 \end{bmatrix} \quad (5)$$

where $l_N = [1,1,\cdots,1]^T$, $\alpha = [\alpha_1,\alpha_2,\cdots,\alpha_N]^T$ and $y = [y_1, y_2,\cdots,y_N]^T$. Ω is a square matrix with entries $\Omega_{ij} = \psi(x_i,y_j) = \phi(x_i)^T\phi(x_i)$, $i,j = 1,\cdots,N$. $\psi(\cdot)$ is a kernel function which satisfies the Mercer's theorem. α and b can be uniquely obtained by solving equation (5), and the linear regression model can be represented by the following formula:

$$y(x) = \sum_{i=1}^{N} \alpha_i \psi(x,x_i) + b \quad (6)$$

In this research, we chose Gaussian kernel function as ψ, so the expression of ψ is expressed as following,

$$\psi(x,y) = \exp(\frac{-\|x-y\|^2}{2\sigma^2}) \quad (7)$$

Finally, the resulting LS−SVM model for function estimation becomes [24],

$$f(x) = \sum_{i=1}^{N} y_i \alpha_i \psi(x,x_i) \quad (8)$$

And the final decision function becomes $\text{sgn}[f(x)]$, where

$$\text{sgn}(x) = \begin{cases} 1, x > 0 \\ -1, x < 0 \end{cases}.$$

As for the situation in this work, the mean vector of the involved process can be used as the input data x_i, and meanwhile, the normal/abnormal patterns of the process can be marked with y_i. Taking the bivariate process as a example, its four classes of normal/abnormal process patterns (i. e. , (0,0), (0,1), (1,0), (1,1)), can be represented with, $y_1 = 1, y_2 = 2, y_3 = 3, y_4 = 4$ respectively, where (0,0) indicate the normal pattern of the mean vector of the bivariate process.

4.2　Design of Multi-class LS-SVM Pattern Recognizer

LS-SVM technique can only construct pattern recognizer

used for two-class datasets. As for the K-class classification problem ($K = 2^p$ in this paper), the "one against one" [25] classification project is used for constructing a K-class pattern recognizer. This project constructs C_K^2 two-class LS-SVM recognizers where each one is trained on dataset from two different classes. As for a test sample X_i, it is inputted into the C_K^2 recognizers in turn to decide which class it belongs to according to the so-called 'Max Wins' voting strategy. In some practical applications, it is possible that very small test data can not be determined which class they belongs to in the light of "one against one" project, this is the so-call "decision blackout area" problem. In this research, some indices (e. g., accuracy rate, missing rate, and false rate) are used to generally evaluate the performance of the K-class pattern recognizers.

5 Model Application-An Industrial Case Study

In this research, we focus on using the proposed model to monitor and diagnose abnormal variable(s) occurred in bivariate process. Among various abnormal conditions, this study is concerned with mean shifts. In this section, we use a case of bivariate process to illustrate the way that the proposed model works.

5.1 Method of Generating Training Datasets

The function of the proposed model is to identify the abnormal variable(s) of the process mean vector by the way of pattern recognition when the T^2 statistic issues an alarm. In this study, the traditional T^2 statistic will work as an abnormal mean shift detector and we formulate the interpretation of the abnormal signal as a pattern recognition problem. The representation of the training datasets has a strong influence on the performance of the LS-

SVM pattern recognizer. Only when the LS-SVM pattern recognizer is trained through suitable training datasets can it acquires the abilities to recognize the patterns of process mean vector. The best way to get training datasets is to collect various data from real-world manufacturing systems, but it costs too much. In this regard, simulation is an effective and useful alternative to investigate into the potential problems associated with the normal/abnormal patterns in a multivariate manufacturing system [26]. We hereby take bivariate process as an example to illustrate the method of generating proper training datasets through simulation. In this research, the sample size of training datasets of each pattern is equal.

Let $(\mu_1, \mu_2)^T$ and $\sum = \begin{pmatrix} \sigma_1^2 & \rho\sigma_1\sigma_2 \\ \rho\sigma_1\sigma_2 & \sigma_2^2 \end{pmatrix}$ denote the mean vector and covariance matrix of a bivariate process, respectively. Let M denotes the sample size of the each class of training datasets, and M should be an integral multiple of 4; Besides, let K_{11}, K_{12}, K_{21} and K_{22} be random number, with K_{11} and $K_{12} \in [-K_1, K_1]$, K_{31} and $K_{32} \in [-K_3, K_3]$, K_{21} and $K_{22} \in [K_2, K_3]$, where $K_1 = 0.3, K_2 = 1.5, K_3 = 3$. In order to generate the proper training datasets in this experiment, we propose the following procedure:

Step1: Employ formula $(\mu_1 + K_{31}\sigma_1, \mu_2 + K_{32}\sigma_2)^T$ to generate the M mean vectors as the training datasets which correspond to the normal pattern of the bivariate process mean vector, and the corresponding class of the normal pattern is marked as (0,0), Here, T^2 statistic is used as a constraint to guarantee that the generated M data (i.e., process mean vectors) make the process in normal state.

Step2: Employ formula $(\mu_1 + K_{11}\sigma_1, \mu_2 + K_{21}\sigma_2)^T$ and $(\mu_1 + K_{12}\sigma_1, \mu_2 - K_{22}\sigma_2)^T$ to generate $M/2$ mean vectors as the training datasets respectively, all of them correspond to the abnormal process pattern of (0,1) class. $K_{11}\sigma_1$ and $K_{12}\sigma_1$ both denote the variations coming from the stochastic disturbance when the process is in the state of normality; Meanwhile, $K_{21}\sigma_2$ and $K_{22}\sigma_2$ both denote the abnormal variation of the process mean vector. In this step, T^2 statistic is also used as a constraint to guarantee that the generated M data make the process in abnormal state.

Step3: Employ formula $(\mu_1 + K_{21}\sigma_1, \mu_2 + K_{11}\sigma_2)^T$ and $(\mu_1 - K_{22}\sigma_1, \mu_2 + K_{12}\sigma_2)^T$ to generate $M/2$ mean vectors as the training datasets respectively, all of them correspond to the abnormal process pattern of (1,0) class. $K_{11}\sigma_2$ and $K_{12}\sigma_2$ both denote the variations coming from the stochastic disturbance when the process is in the state of normality; Meanwhile, $K_{21}\sigma_1$ and $K_{22}\sigma_1$ both denote the abnormal variation of the process mean vector. And T^2 statistic is also used as a constraint to guarantee that the generated M data make the process in abnormal state.

Step4: Employ formula $(\mu_1 + K_{21}\sigma_1, \mu_2 + K_{22}\sigma_2)^T$, $(\mu_1 + K_{21}\sigma_1, \mu_2 - K_{22}\sigma_2)^T$, $(\mu_1 - K_{21}\sigma_1, \mu_2 + K_{22}\sigma_2)^T$, and $(\mu_1 - K_{21}\sigma_1, \mu_2 - K_{22}\sigma_2)^T$ to generate $M/4$ mean vectors as the training datasets respectively, all of them correspond to the abnormal process pattern of (1,1) class. And T^2 statistic is also used as a constraint to guarantee that the generated M data make the process in abnormal state.

At each step of the procedure, the T^2 statistic is used as a constraint to guarantee that the generated data meet the demand of the corresponding patterns of the process mean vector. At the end of the subsequent experiment, we need to test the performance of the designed pattern recognizer. The same procedure is employed to generate the test datasets.

5.2 Case Analysis and Discussion

5.2.1 Training Dataset

In this research, a bivariate industrial case derived from Montgomery (1991) [27] is provided to explain how the proposed LS-SVM pattern recognition model works. This case was also studied by literature [28] (Ma Yizhong, 2004) to illustrate his PCA method of process diagnosis. In this case, the two variables X1 and X2 of a bivariate process represent process target mean values from a chemical plant, and the reference mean vector $\overline{X}$ and the reference variance/covariance matrix S can be estimated from the initial fifteen observations (n=15),

$$\overline{X} = [10 \quad 10]^T, S = \begin{bmatrix} 0.7986 & 0.6793 \\ 0.6793 & 0.7343 \end{bmatrix}$$

Let $\alpha=0.05$, and then the control limits of T^2 chart can be calculated as following according to formula (2) and (3),

$$UCL_{T^2} = \frac{2 \times 16 \times 14}{15 \times 13} \times 3.81 = 8.753, LCL_{T^2} = 0$$

Due to extensive demands for training datasets in our experiment, Monte-Carlo simulation was used to generate training datasets. The simulation was implemented using MATLAB7 software to perform the diagnosis procedure, and MATLAB M-files were developed. Using the procedure in section 5.1 and control limits obtained above, we obtain a group of simulated training data (i.e., mean vector $X = (X_1 - X_2)^T$) at different parameters. As an illustration, Table 1 summarizes the resulting values of the training datasets when the sample size is 24 ($M=24$) for each of the four patterns of the process mean vector. The T^2 values of the training dataset are also presented in Table 1 to indicate that if the training data cause an abnormal state in the bivariate process.

Tab 1 A group of training datasets for patterns of bivariate process mean vector and their T² statistics values ($M=24, K_1=0.3, K_2=1, K_3=3$)

Pattern (0,0) training data	T²	Pattern (0,1) training data	T²	Pattern (1,0) training data	T²	Pattern (1,1) training data	T²
$(10.96,10.14)^T$	4.05	$(10.11,12.04)^T$	24.27	$(12.56,9.68)^T$	48.08	$(12.63,12.06)^T$	8.85
$(9.13,9.11)^T$	1.09	$(10.23,12.03)^T$	21.43	$(12.45,9.99)^T$	35.4687	$(10.91,12.51)^T$	20.43
$(9.42,8.49)^T$	6.97	$(9.64,12.01)^T$	34.33	$(12.15,9.78)^T$	32.59	$(11.04,12.44)^T$	16.87
$(10.57,11.01)^T$	2.14	$(10.36,11.88)^T$	16.06	$(12.59,10.39)^T$	29.19	$(12.63,12.40)^T$	8.82
$(10.96,11.90)^T$	8.65	$(9.66,12.39)^T$	46.28	$(11.51,9.95)^T$	14.29	$(12.68,11.17)^T$	16.82
$(11.97,11.44)^T$	5.21	$(9.65,12.19)^T$	39.78	$(12.17,9.88)^T$	30.41	$(12.55,11.30)^T$	12.95
$(9.10,9.51)^T$	1.51	$(10.31,12.16)^T$	23.10	$(12.59,9.97)^T$	40.44	$(12.50,8.89)^T$	75.04
$(12.25,11.43)^T$	7.84	$(9.62,12.25)^T$	42.44	$(11.80,9.92)^T$	20.59	$(11.41,8.61)^T$	45.32
$(10.95,10.58)^T$	1.43	$(10.32,12.23)^T$	24.57	$(11.98,10.37)^T$	15.92	$(12.02,8.28)^T$	80.48
$(10.11,9.83)^T$	0.47	$(9.59,11.83)^T$	30.59	$(11.75,10.24)^T$	13.71	$(11.15,7.98)^T$	58.98
$(12.03,11.26)^T$	6.50	$(10.12,12.52)^T$	37.52	$(11.30,9.85)^T$	12.16	$(11.52,8.91)^T$	39.25
$(9.35,8.79)^T$	3.30	$(9.59,11.90)^T$	32.60	$(11.85,9.51)^T$	31.73	$(12.39,9.08)^T$	62.70
$(11.66,11.11)^T$	4.04	$(10.49,7.90)^T$	40.68	$(8.50,9.89)^T$	11.58	$(8.18,11.00)^T$	83.91
$(9.47,10.11)^T$	2.34	$(10.03,7.49)^T$	41.12	$(7.33,9.94)^T$	40.14	$(7.90,12.21)^T$	107.54
$(11.77,11.02)^T$	5.45	$(9.58,7.87)^T$	20.21	$(8.18,9.62)^T$	12.87	$(7.82,12.03)^T$	102.33
$(8.44,9.34)^T$	5.83	$(10.36,8.97)^T$	11.53	$(7.41,9.84)^T$	35.09	$(8.03,11.79)^T$	81.25
$(8.56,8.79)^T$	2.59	$(10.47,7.49)^T$	54.10	$(9.04,10.32)^T$	9.45	$(8.72,11.28)^T$	37.89
$(12.06,12.11)^T$	6.11	$(10.02,8.70)^T$	11.04	$(7.85,10.42)^T$	38.05	$(8.92,12.29)^T$	67.08
$(11.66,11.80)^T$	4.40	$(10.09,8.68)^T$	12.43	$(8.56,10.01)^T$	12.41	$(8.29,7.61)^T$	9.27
$(8.52,8.05)^T$	5.80	$(9.79,8.63)^T$	9.20	$(7.74,9.64)^T$	22.14	$(7.48,9.12)^T$	18.14
$(7.84,8.26)^T$	5.88	$(10.50,8.77)^T$	17.79	$(7.37,9.69)^T$	32.34	$(9.02,7.75)^T$	14.04
$(8.83,9.98)^T$	7.67	$(10.19,7.76)^T$	36.90	$(7.62,9.60)^T$	24.03	$(8.93,7.82)^T$	11.79
$(11.27,10.75)^T$	2.67	$(10.47,8.35)^T$	27.16	$(7.34,9.73)^T$	34.19	$(7.58,9.06)^T$	15.40
$(9.51,9.13)^T$	1.59	$(9.82,8.59)^T$	10.10	$(7.34,9.67)^T$	32.76	$(7.34,8.07)^T$	9.56

According to Table 1, it is easy to verify that the T^2 values of training data for normal pattern totally do not exceed the control limit UCL_{T^2} (8.753), and meanwhile the T^2 values of training data for abnormal patterns totally exceed the control limit UCL_{T^2} (8.753).

For the purpose of comparing the effect of the training sample size on the performance of the diagnosis model, we did the experiment under different training sample size M, with $M=8,16,24,36,44$, and 52 increasingly. According to probability and statistic, we consider that the training sample size M is smaller when $M=8,16,24$; while, the training sample size is bigger when $M=36,44,52$. Furthermore, the experiment was also done under smaller normal disturbance and bigger stochastic disturbance of mean vector, which correspond to $K_1=0.3$ and $K_1=0.6$ respectively. Similarly, for an illustrative purpose, some of the distribution maps of the training data are shown in Figure 2 below. The four kinds of training datasets, which correspond to the patterns (0,0), (0,1), (1,0) and (1,1), are marked with circle (○), asterisk (*), X-mark(×) and triangle (▽) on each of the distribution maps.

According to the distribution maps of training datasets and the SVM theory, it is observed that the four defined patterns of the training datasets can be correctly classified by the optimal separating hyper plane in the high dimensional feature space. From statistics, when the T^2 restrict condition is employed to generate the four patterns, the data that belongs to the normal pattern is inside of a elliptical area on the distribution map, and meanwhile the data that belongs to the three abnormal patterns is outside of the elliptical area on the distribution map. We can see this clearly from Figure 2 especially when the sample size M is bigger.

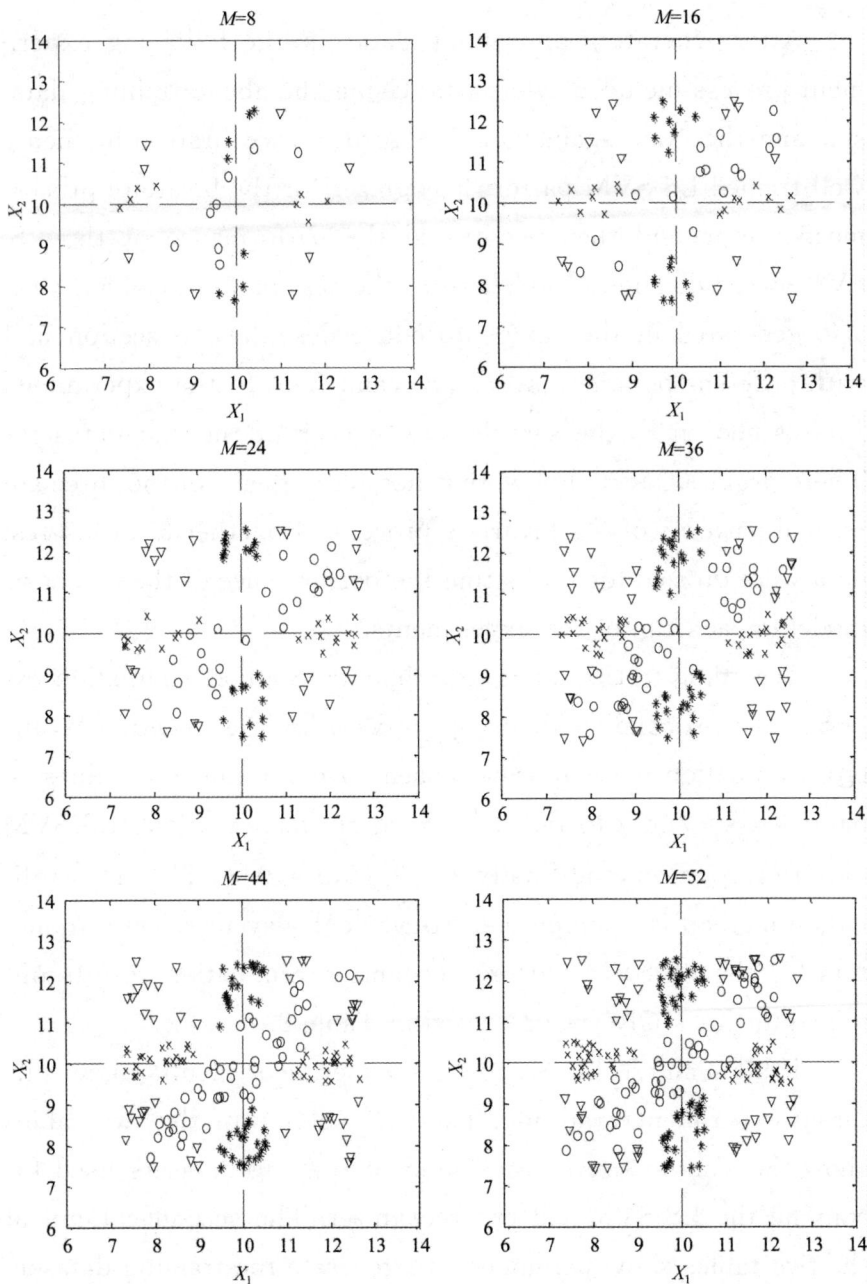

Fig 2　Distribution maps of Training datasets （$K_1=0.3, K_2=1, K_3=3$）

5.2.2 Results and Discussion

After generating of training datasets, the following experiment process included two parts: Using the above training datasets and the "one against one" algorithm, we firstly obtained a well-trained LS-SVM pattern recognizer for the bivariate process mean vector; and then, we tested the performance of the LS-SVM-based diagnosis model using the test datasets, which were also generated in the same procedure described in section 5.1 with different parameters (K_1, K_2, and K_3). In the experiment, 1000 is chosen as the sample size of each of the test datasets. Therefore, 1000 test data were generated for each of the three abnormal patterns of the bivariate process, altogether 3×1000 test data were utilized for evaluating the performance of the diagnosis model in each run of the experiments.

According to the above experimental method, simulation experiment was conducted. In the process of the experiment, the cross-validation method was chosen to determine the values of parameters σ and γ to improve the performance of the LS-SVM pattern recognizer, and finally $\gamma = 0.15$, $\sigma = 0.3$. The cross-validation method is a simple and convenient way to select parameters for the LS-SVM pattern recognizer. Simulation results obtained in this study are presented in Table 2.

Table 2 and Table 3 show the experiment results under different experiment condition. The first column of the two tables show the sample size of each kind of training datasets used for training the LS-SVM pattern recognizer. The second column of the two tables show parameters to generate the training datasets according to the procedure introduced in section 5.1. The third column of the two tables show the categories of each 1000 test dataset, which are used for testing the performance of the diagnosis model. The second row of the third column in the two tables indicates the parameters to generate these test datasets. Columns 4—7 in the two tables show the number of times that the 1000 test data reached the corresponding classified patterns. Columns 8—10 provide three performance indexes (i. e. , accuracy rate, false rate, and missing rate) to evaluate the diagnosis model as a whole. We note that the difference between Table 2 and Table 3 is the value of parameter K_1 in the third column of the two tables. In this research, we indicate smaller common cause variation coming from the stochastic disturbance of the process with $K_1 = 0.3$ in the third column of Table 2; While, we indicate bigger common cause variation coming from the stochastic disturbance of the process with $K_1 = 0.6$ in the third column of Table 3.

Tab 2 Performance of the LS-SVM pattern recognizer for bivariate process with smalll disturbance

Parameters for training datasets	Parameters for 1000 test data	Conditions of pattern classification				Classification percentage (%)		
M (K_1, K_2, K_3)	$(K_1=0.3, K_2=1, K_3=3)$	(0,0)	(0,1)	(1,0)	(1,1)	Accuracy rate	False Rate	Missing rate
8 (0.3,1,3)	(0,1)	16	984	0	0	98.4	0	1.6
	(1,0)	5	0	993	2	99.3	0.2	0.5
	(1,1)	168	7	39	786	78.6	4.6	16.8
(0.6,1,3)	(0,1)	33	967	0	0	96.7	0	3.3
	(1,0)	0	0	1000	0	100	0	0
	(1,1)	78	105	5	812	81.2	11	7.8

续表

M	Parameters for training datasets (K_1, K_2, K_3)	Parameters for 1000 test data $(K_1=0.3, K_2=1, K_3=3)$	\multicolumn{4}{c	}{Conditions of pattern classification}	\multicolumn{3}{c	}{Classification percentage (%)}
			(0,0)	(0,1)	(1,0)	(1,1)	Accuracy rate	False Rate	Missing rate
16	(0.3,1,3)	(0,1)	0	1000	0	0	100	0	0
		(1,0)	0	0	1000	0	100	0	0
		(1,1)	14	3	39	944	94.4	4.2	1.4
	(0.6,1,3)	(0,1)	0	1000	0	0	100	0	0
		(1,0)	0	0	998	2	99.8	0.2	0
		(1,1)	46	32	1	921	92.1	3.3	4.6
24	(0.3,1,3)	(0,1)	0	1000	0	0	100	0	0
		(1,0)	0	0	1000	0	100	0	0
		(1,1)	43	6	9	942	94.2	1.5	4.3
	(0.6,1,3)	(0,1)	0	1000	0	0	100	0	0
		(1,0)	18	0	982	0	98.2	0	1.8
		(1,1)	15	18	23	944	94.4	4.1	1.5
36	(0.3,1,3)	(0,1)	0	1000	0	0	100	0	0
		(1,0)	0	0	1000	0	100	0	0
		(1,1)	25	0	0	975	97.5	0	2.5
	(0.6,1,3)	(0,1)	0	1000	0	0	100	0	0
		(1,0)	0	0	1000	0	100	0	0
		(1,1)	12	7	17	964	96.4	2.4	1.2
44	(0.3,1,3)	(0,1)	0	1000	0	0	100	0	0
		(1,0)	0	0	1000	0	100	0	0
		(1,1)	9	1	1	989	98.9	0.2	0.9
	(0.6,1,3)	(0,1)	0	1000	0	0	100	0	0
		(1,0)	0	0	1000	0	100	0	0
		(1,1)	1	1	19	979	97.9	2	0.1

续表

Parameters for training datasets		Parameters for 1000 test data	Conditions of pattern classification				Classification percentage (%)		
M	(K_1,K_2,K_3)	$(K_1=0.3, K_2=1, K_3=3)$	(0,0)	(0,1)	(1,0)	(1,1)	Accuracy rate	False Rate	Missing rate
52	(0.3,1,3)	(0,1)	0	1000	0	0	100	0	0
		(1,0)	0	0	1000	0	100	0	0
		(1,1)	10	0	2	988	98.8	0.2	1
	(0.6,1,3)	(0,1)	0	1000	0	0	100	0	0
		(1,0)	0	0	1000	0	100	0	0
		(1,1)	0	4	19	977	97.7	2.3	0

Tab 3 Performance of the LS-SVM pattern recognizer for bivariate process with big disturbance

Parameters for training datasets		Parameters for 1000 test data	Conditions of pattern classification				Classification percentage (%)		
M	(K_1,K_2,K_3)	$(K_1=0.3, K_2=1, K_3=3)$	(0,0)	(0,1)	(1,0)	(1,1)	Accuracy rate	False Rate	Missing rate
8	(0.3,1,3)	(0,1)	45	906	0	49	90.6	4.9	4.5
		(1,0)	0	0	1000	0	100	0	0
		(1,1)	101	36	143	720	72	17.9	10.1
	(0.6,1,3)	(0,1)	0	1000	0	0	100	0	0
		(1,0)	13	0	975	12	97.5	1.2	1.3
		(1,1)	27	60	20	893	89.3	8	2.7
16	(0.3,1,3)	(0,1)	2	998	0	0	99.8	0	0.2
		(1,0)	42	0	958	0	95.8	0	4.2
		(1,1)	35	38	19	908	90.8	5.7	3.5
	(0.6,1,3)	(0,1)	0	1000	0	0	100	0	0
		(1,0)	0	0	984	16	98.4	1.6	0
		(1,1)	12	56	28	904	90.4	8.4	1.2

续表

Parameters for training datasets		Parameters for 1000 test data	Conditions of pattern classification				Classification percentage (%)		
M	(K_1, K_2, K_3)	$(K_1=0.3, K_2=1, K_3=3)$	(0,0)	(0,1)	(1,0)	(1,1)	Accuracy rate	False Rate	Missing rate
24	(0.3,1,3)	(0,1)	14	986	0	0	98.6	0	1.4
		(1,0)	1	0	999	0	99.9	0	0.1
		(1,1)	18	12	23	947	94.7	3.5	1.8
	(0.6,1,3)	(0,1)	0	997	0	3	99.7	0.3	0
		(1,0)	0	0	1000	0	100	0	0
		(1,1)	43	6	3	948	94.8	0.9	4.3
36	(0.3,1,3)	(0,1)	7	977	0	16	97.7	1.6	0.7
		(1,0)	0	0	997	3	99.7	0.3	0
		(1,1)	26	2	5	967	96.7	0.7	2.6
	(0.6,1,3)	(0,1)	2	998	0	0	99.8	0	0.2
		(1,0)	0	0	1000	0	100	0	0
		(1,1)	15	9	12	964	96.4	2.1	1.5
44	(0.3,1,3)	(0,1)	0	1000	0	0	100	0	0
		(1,0)	0	0	1000	0	100	0	0
		(1,1)	16	0	0	984	98.4	0	1.6
	(0.6,1,3)	(0,1)	0	1000	0	0	100	0	0
		(1,0)	0	0	1000	0	100	0	0
		(1,1)	20	0	15	965	96.5	1.5	2
52	(0.3,1,3)	(0,1)	0	1000	0	0	100	0	0
		(1,0)	0	0	1000	0	100	0	0
		(1,1)	0	0	10	990	99	1	0
	(0.6,1,3)	(0,1)	0	1000	0	0	100	0	0
		(1,0)	0	0	1000	0	100	0	0
		(1,1)	0	0	2	998	99.8	0.2	0

From Table 2, Table 3 and their comparison, we can see the experimental results below.

1) Sample size of the training datasets: Increasing the sample size of the training datasets can improve the accuracy rate of the diagnosis model. But the accuracy rates of each of the abnormal patterns can attain more than 90% even under the conditions of 16 or 24 training sample size. So we think the diagnosis model can obtain good performance of accuracy rate under the condition of small sample size (less than 30 in terms of statistic theory). On the other hand, it is difficult to obviously improve the performance of the diagnosis model under the condition of large sample size (more than 30 in terms of statistic theory) with expensive cost.

2) Effect of stochastic disturbance on the performance of the diagnosis model: first, when the stochastic disturbance in the training datasets is increased (i.e., the K_1 value of the training datasets is changed from 0.3 to 0.6, with other experiment parameters held unchanged), we can obtain almost the same performance of accuracy rate. Second, when the stochastic disturbance in the test datasets is increased (i.e., the K_1 value of the test datasets is changed from 0.3 to 0.6, with other experiment parameters held unchanged), the accuracy rates of each of the abnormal patterns just decrease a little, the effect of the stochastic disturbance in test datasets can be neglected. So the diagnosis model possesses good capability of disturbance-resistance.

3) The two parameters (γ, σ) have significant effect on the performance of the LS-SVM pattern recognizer, and they must be chosen reasonably. The cross-validation method is simply employed to choose the two parameters in the experiment. In order to further improve the performance of the accuracy rate, especial-

ly under the condition of small training sample size, some intelligent algorithm can be introduced in this research to optimize the two parameters.

We can notice an additional phenomenon in the experiment, that is, the accuracy rate of (1,1) pattern is always smaller than that of (0,1) or (1,0) pattern under the same experimental conditions.

5.2.3 Comparison

Jackson (1991) [29] also studied this industrial case, and further pointed out that the four observations of process mean vector $A=(12.3,12.5)^T, B=(7.0,7.3)^T, C=(11.0,9.0)^T$, and $D=(7.3,9.1)^T$ were all in the state of abnormality. In this research, the patterns of the above four mean vectors was diagnosed with (1,1), (1,1), (1,1), and (1,1) respectively by our proposed diagnosis model under the experiment conditions of $M=52, K_1=0.6, K_2=1, K_3=3$. The diagnosis results is just the same as Jackson's, but still, we can obtain some more detailed information about the causes of the process abnormality from our diagnosis result. In the light of the above single instance, though possibly it is not perfect and comprehensive enough, we consider that the proposed diagnosis model in this research is able to provide more information about the causes of the process abnormality.

6 Conclusions and Recommendations for Future Work

In today's multivariate manufacturing scenario, identifying source(s) of abnormal signals has been a challenging task for traditional MSPC techniques. Lots of artificial intelligence techniques (e.g., ANN, SVM) have been suggested as alternatives to MSPC charts because of their superior performance. This research presented a new LS-SVM based model to classify and identify the abnormal variable(s) of the process mean in the bivariate process. In the proposed model, T^2 statistic (or other multivariate control charts) are used as the detector of the overall process abnormality, and a LS-SVM ensemble approach is developed for identifying the abnormal variable(s) of abnormal signals in bivariate processes. As a result of LS-SVM technique, another feature of this model is that the model can meet the above requirements under the condition of small sample size.

The performance of the model for classification and identification rates of abnormal mean shifts that appear in bivariate process is evaluated through the use of a numerical example from chemical industry. The experiment on bivariate process demonstrates the effectiveness of the proposed model, and the results of the model implementation are inspiring and promising. We believe that the proposed LS-SVM based diagnosis model is a good and strong method used for identifying the abnormal variable(s) in bivariate quality control.

In the process of the experiment, we note that the parameters of LS-SVM have significant effect on the performance of LS-SVM based diagnosis model. In order to improve the performance of the diagnosis model, especially on the condition of the small sample size of the training datasets, We suggest introducing intelligent evolutionary algorithm (e.g., Particle Swarm Optimization, PSO) to optimize the parameters of LS-SVM recognizer in the future work. Although this research mainly considers the situation of the bivariate process, the proposed approach can be employed in other aspects, especially to determine the magnitude of a shift, or potentially to be extended to multivariate SPC in general.

Acknowledgement

This work was supported by two National Natural Science Foundation of China, Grant No. 70931002 and No. 70672088.

References

[1] E. N. Loredo, D. Jearkpaporn, C. M. Bororo, "Model-based control chart for autoregressive and correlated Data", Quality and Reliability Engineering International, 2002, Vol. 18, pp. 489−496.

[2] Aparisi F, Haro C. L., "Hotelling's T^2 control chart with variable sampling intervals", International Journal of Production Research, Vol. 39, pp. 3127−3140, 2001.

[3] Lowry CA, Montgomery DC, "A review of multivariate control charts", IIE Transactions, 1995, Vol. 27, pp. 800−810.

[4] Mason R. L, Champ C. W., Tracy N. D., Wierda S. J., Young J. C., "Assessment of multivariate process control techniques", Journal of Quality Technology, 1997, Vol. 29, pp. 140−143.

[5] R. B. Crosier, "Multivariate generalizations of cumulative sum quality control schemes", Technometrics, 1988, Vol. 30, pp. 291−303.

[6] S. S. Prabhu, G. C. Runger, "Designing a multivariate EWMA control chart", Journal of Quality Technology, 1997, Vol. 29, pp. 8−15.

[7] Louwerse D. J., Smilde A. K., "Multivariate statistical process control of batch processes based on three-way models", Chemical Engineering Science, 2000, Vol. 55, pp. 1225−1235.

[8] Nomikos P, MacGregor J. F., "Multi-way partial least squares in monitoring batch processes", Chemometrics and Intel-

ligent Laboratory Systems,1995,Vol. 30,pp. 97—108.

[9] S. Bersimis, S. Psarakis, J. Panaretos, "Multivariate statistical process control charts: An overview", Quality and Reliability Engineering International,2007,Vol. 23,pp. 517—543.

[10] Alt F. B, Multivariate quality control, Encyclopedia of the Statistical Science, Johnson NL, Kotz S, Read CR (eds.). Wiley,New York,1985,Vol. 6,pp. 111—122.

[11] R. L. Mason, J. C. Young, "Improving the sensitivity of the T^2 statistic in multivariate process control", Journal of Quality Technology,1999,Vol. 31,pp. 155—165.

[12] G. C. Runger, F. B. Alt, D. C. Montgomery, "Contributors to a multivariate statistical process control signal", Communications in Statistics-Theory and Methods, 1996, Vol. 25, pp. 2203—2213.

[13] Ho ES, Chang SI, "An integrated neural network approach for simultaneous monitoring of process mean and variance shifts—a comparative study", International Journal of Production Research,1999,Vol. 37,no. 8,pp. 1881—1901.

[14] Noorossana R, Farrokhi M. Saghaei A, "Using neural networks to detect and classify out-of-control signals in autocorrelated processes", Quality and Reliability Engineering International,2003,Vol. 19,pp. 1—12.

[15] Seyed Taghi Akhavan Niaki, Babak Abbasi, "Fault Diagnosis in Multivariate Control Charts Using Artificial Neural Networks", Quality and Reliability Engineering International, 2005,Vol. 21,pp. 825—840.

[16] R. S. Guh, "On-line Identification and Quantification of Mean Shifts in Bivariate Processes Using a Neural Network-based Approach", Quality and Reliability Engineering International,2007,Vol. 23,pp. 367—385.

[17] Jian-bo Yu, Li-feng Xi, "A neural network ensemble-based model for on-line monitoring and diagnosis of out-of-control signals in multivariate manufacturing processes", Expert Systems with Applications, 2009, Vol. 36, pp. 909—921.

[18] Shih-Yen Lin, Young-Jou Lai, Shing I Chang, "Short-run statistical process control: multicriteria part family formation", Quality and Reliability Engineering International, 1997, Vol. 13, no. 1, pp. 9—24.

[19] J. L. Alfaro, J. F. Ortega, "A Robust Alternative to Hotelling's T^2 Control Chart Using Trimmed Estimators", Quality and Reliability Engineering International, 2008, Vol. 24, pp. 601—611.

[20] V. N. Vapnik, The nature of statistical learning theory, Springer, New York, 1995.

[21] Nello Cristianini, John Shawe-Taylor, An Introduction to Support Vector Machines and Other Kernel-based Learning Methods, Cambridge University Press, 2005.

[22] J. A. K. Suykens, J. Vandewalle, "Least Squares Support Vector Machine Classifiers", Neural Processing Letters, 1999, Vol. 9, pp. 293—300.

[23] Suykens Johan A K, Gestel Tony Van, Brabanter Jos De, Least Squares Support Vector Machines, World Scientific Publishers, Singapore, 2002.

[24] C. L. Huang, C. J. Wang, "A GA-based feature selection and parameters optimization for support vector machines", Expert Systems with Applications, 2006, Vol. 31, pp. 231—240.

[25] Bo Liu, Zhifeng Hao, Xiaowei Yang, "Nesting algorithm for multi-classification problems", Soft Computing, 2007, Vol. 11, pp. 383—389.

[26] F. Zorriassatine, J. D. T. Tannock, C. O. Brien, "Using

novelty detection to identify abnormalities caused by mean shifts in bivariate processes", Computers & Industrial Engineering, 2003, Vol. 44, pp. 385—408.

[27] D. C. Montgomery, Introduction to Statistical Quality Control, John Wiley & Sons, New York, 1991.

[28] MaYizhong, Theory and Methods of Reducing and Controlling Variation of Multivariate Quality Characteristics, Northwestern Polytechnical University, China, 2002.

[29] J. E. Jackson, A User's Guide to Principle Components, John Wiley & Sons, New York, 1991.

Variance Shifts Identification Model of Bivariate Process Based on LS-SVM Pattern Recognizer

Cheng Zhi-Qiang[1], Ma Yi-Zhong[1], Bu Jing[2]

[1]*Department of Management Science and Engineering, Nanjing University of Science and Technology, Nanjing Jiangsu, 210094, P. R. China.*

czqme@163.com, yzma-2004@163.com

[2]*Automation Institute, Nanjing University of Science and Technology, Nanjing Jiangsu, 210094, P. R. China.*

bujing30@foxmail.com

Abstract: Multivariate Statistical Process Control (MSPC) techniques are effective tools for detecting the abnormalities of multivariate process variation. MSPC techniques are based on overall statistics, this has caused the difficulties in interpretation of the alarm signal, that is, MSPC charts do not provide the necessary information about which process variables (or subset of them) are responsible for the signal, and this task is left up to the quality engineers in production field. This article proposes a model based on LS-SVM pattern recognizer to diagnose the bivariate process abnormality in covariance matrix. The main property of this model is a supplement of MSPC |S| chart to identify the variable(s) which is (are) responsible for the process abnormality when |S| chart issue a warning signal. Through simulation experiment, the performance of the model is evaluated by accuracy rate of pattern recognition. The results indicate that the proposed model is an effective method to interpret the root causes of the process abnormality. A bivariate example is presented to illustrate the application of the proposed model.

Key words: complex products; multivariate statistical control; quality diagnosis; least squares support vector machine; pattern recognition

1 Introduction

In many complex products manufacturing settings, the manufacturing process may have two or more correlated quality characteristics which must be controlled simultaneously. Many quality researchers have proposed different multivariate control charts to monitor the multivariate process abnormalities (out of control states). The most common abnormalities of quality characteristics are mean shifts and variance shifts. This situation results in the most rapid development of MSPC [1]. Most of the previous studies are focused on diagnosing the process abnormality through multivariate control charts. There already exist a variety of techniques used for monitoring the mean vector of multivariate process. These techniques traditionally include the well-known Hotelling's T^2 charts [2], multivariate cumulative sum (MCUSUM) [3], multivariate exponentially weighted moving average (MEWMA) [4], and their improved methods [5]. Another kind of study is concerned with variance shifts. Montogomery [1] presented the log $|s|$ control charts to monitor variances of multivariate process. Alt [6] introduced the generalized variance $|S|$ statistic to control the process variances. Machado and Costa et al. [7] proposed VMAX statistic to particularly monitor the abnormalities of covariance matrix of bivariate processes. These MSPC tools are still widely used in the area of the diagnosis of process abnormalities through overall statistics of the process.

But traditional MSPC techniques have an obvious defect, that is, they can not interpret and tell which variable or combination of variables caused the out-of-control signals. On the other hands, it is very important for the quality engineers to rapidly identify the variable(s) which is (are) influenced by the assignable causes and make some adjustments to bring the process back to the in-control condition.

In recent years, many quality practitioners have made great efforts to develop alternative techniques for process diagnosis. Typical examples of such techniques are machine learning methods, for example, Artificial Neural Networks (ANN) and Support Vector Machines (SVM). Many researches have been working on monitoring the mean shifts and variance shifts in multivariate process by using ANN methods [8,9,10]. However, the training of ANN needs so many samples that it is not always feasible when ANN is used in a real application. It is necessary for quality practitioners to study the process diagnosis techniques which can use as less samples as possible. Support Vector Machine [11] is quite an efficient pattern recognition technique developed on statistical learning theory, its effective application can be found in the area of solving problems of small training samples, nonlinear and high dimension. And least squares support vector machine (LS-SVM) [12] is an extension of standard SVM with faster computational speed.

The aim of this research is to propose a fault diagnosis model used for variance shifts in bivariate process using LS-SVM techniques. This model is applied not only to monitor the abnormal changes of covariance matrix but also to identify the variable(s) which is (are) responsible for the abnormalities. In the mean time, this intelligent model can work with less training samples than ANN techniques. As a result of the above-mentioned considerations, this research proposes a bivariate process diagnosis model based on the combination of traditional |S| chart method and LS-SVM pattern recognition technique, with focus on the abnormal variance shifts in bivariate process. The performance of the proposed model is evaluated and compared under different experimental conditions. An example is given to demonstrate the feasibility and validity of the model.

The rest of this paper is structured as follows: Section 2 presents a LS-SVM recognizer based model for identifying the assignable causes in bivariate processes. Section 3 briefly introduces the |S| chart and the definition of covariance matrix patterns. Section 4 describes LS-SVM and its working principle. Section 5 presents the results of performance evaluation. Section 6 presents an application instance of the proposed model.

2 LS-SVM based diagnosis model

MSPC charts are applied in two distinct stages of manufacturing [13]. In stage one, charts are used for testing whether the process is in control and establishing the standard of the in-control process statistically. In stage two, charts are used for testing whether the process remains in control with the known process parameters (μ_0, Σ_0). In this article, we investigate using LS-SVM based model for monitoring and diagnosing variance shifts of bivariate manufacturing processes in the second stage.

The proposed model includes two parts mainly: the variance shifts detector and the LS-SVM pattern recognizer, schematically shown in Figure 1. In the first part, |S| chart works as a variance shifts detector. The control chart technique may be regarded as the graphical expression and operation of statistical hypothesis testing for detecting the process abnormalities. Once a abnormal alarm is signaled by |S| control chart, the proposed model formulates the interpretation of abnormal signal as a pattern classification problem, and the LS-SVM pattern recognizer in the second part will determine which variable(s) is (are) responsible for the variance shifts.

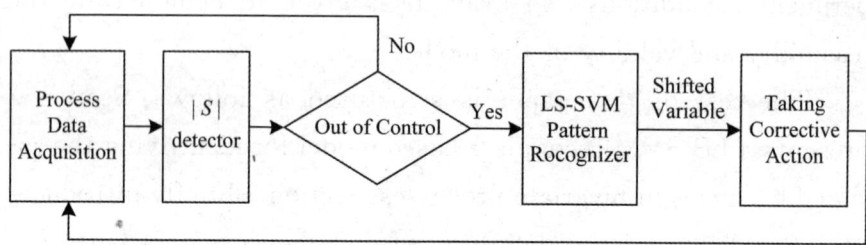

Fig 1   Bivariate Process Variance Shift Diagnosis and Identification Model by LS-SVM Pattern Recognizer

From the model, it can be seen that though pattern recognition method is extremely different from the traditional MSPC tools, however, it can complement and enhance (not replace) the power of the latter [14]. The focus of this model is the interpretation of the process variance shifts by using LS-SVM pattern recognition method.

3 |S| chart and patterns of the covariance matrix in bivariate process

3.1 |S| chart

Variability of multivariate process is usually summarized in $p \times p$ the covariance matrix Σ, where p represents the number of process variables. For a bivariate process, we assume that the two quality characteristics are symbolized with vector $X = (x_1, x_2)$, which follow a bivariate normal distribution with mean vector $\mu_0 = (\mu_{10}, \mu_{20})$ and in-control covariance matrix $\Sigma_0 = \begin{pmatrix} \sigma_1^2 & \rho\sigma_1\sigma_2 \\ \rho\sigma_1\sigma_2 & \sigma_2^2 \end{pmatrix}$, namely, $X \sim N(u_0, \Sigma_0)$, and ρ is the correlation coefficient among variable x_1 and variable x_2. The standard values of μ_0 and Σ_0 are either known or derived from the historical data of the controlled process. Another assumption in this paper is that the values of μ_0 and ρ remain unchanged even if abnormal disturbance occurs in the bivariate process, in other words, this research assumes that the assignable cause does not shift process mean μ_0 and correlation coefficient ρ among x_1 and x_2.

According to the generalized variance $|S|$ chart proposed by Alt [6], for a given subgroup size $n9n > p$), the generalized variance $|S|$ of process sample vectors $X_1, X_2, \cdots, X_n$ is calculated as:

$$|S| = \left| \frac{1}{n-1} \sum_{i=1}^{n} (X_i - \overline{X})(X_i - \overline{X})^{\mathrm{T}} \right| \tag{1}$$

where, $\overline{X} = \frac{1}{n} \sum_{i=1}^{n} X_i$. Represent $E(|S|)$ and $V(|S|)$ for the mean and variance of $|S|$ respectively, then the most of the probability distribution of $|S|$ is contained in the interval $E(|S|) \pm 3\sqrt{V(|S|)}$. It can be shown that, the upper control limit (UCL), central line (CL) and lower control limit (LCL) of $|S|$ chart would be:

$$UCL = |\Sigma_0|(b_1 + 3b_2^{1/2}) \tag{2}$$
$$CL = b_1|\Sigma_0| \tag{3}$$
$$LCL = Max\{0, |\Sigma_0|(b_1 - 3b_2^{1/2})\} \tag{4}$$

Coefficients b_1 and b_2 are calculated as:

$$b_1 = (n-1)^{-p} \prod_{i=1}^{p} (n-i) \tag{5}$$

$$b_2 = (n-1)^{-2p} \prod_{i=1}^{p} (n-i) \left[\prod_{j=1}^{p} (n-j+2) - \prod_{j=1}^{p} (n-j) \right] \tag{6}$$

In practice, let S_j denote the covariance matrix of j th sample of a in-control process, $\overline{S}$ denote the mean of $S_j, j = 1, 2, \cdots, m$, then $|\Sigma_0|$ can be estimated from $|\overline{S}|/b_1$, and $|\overline{S}|$ is the determinant of $\overline{S}$.

3.2 Patterns definition of covariance matrix

When abnormal disturbance affects the process variance, the in-control covariance matrix Σ_0 is changed to Σ_1. Let $d^2 = |\Sigma_1|/|\Sigma_0|$, then d^2 indicates the "determinant ratio, DR" between out-of-control and in-control covariance matrices, the magnitude of the disturbance is determined by d^2. In this research, we investigate three kinds of abnormal patterns caused by abnormal disturbances that change the initial in-control covariance matrix to $\Sigma_1 = \begin{pmatrix} a^2\sigma_1^2 & \rho a b \sigma_1 \sigma_2 \\ \rho a b \sigma_1 \sigma_2 & b^2\sigma_2^2 \end{pmatrix}$. In terms of the expression Σ_1 of bivariate process, the patterns of Σ_1 can be defined. The first abnormal pattern considers that the abnormal disturbance only affects the variance of x_1, that is, $a^2 = d^2 > 1$ and $b = 1$, and this pattern can be represented by a symbol (1,0); the second abnormal pattern considers that the abnormal disturbance only affects the variance of x_2, in this case, $b^2 = d^2 > 1$ and $a = 1$, and this pattern can be represented by a symbol (0,1); the third abnormal pattern considers that abnormal disturbance affects both the variance of x_1 and the variance of x_2, in this case, $a^2 = b^2 = d^2 > 1$, and this pattern can be represented by a symbol (1,1). The fourth pattern is the normal pattern, that is, $a = b = d = 1$, and this pattern can be represented by a symbol (0,0). By classifying the patterns of the covariance matrix, the variable(s) which is (are) responsible for the process abnormalities can be identified accordingly.

4 LS-SVM based pattern recognizer

4.1 LS-SVM principle

Let x_i be a set of data points and y_i indicate the corresponding classes, where $x_i \in R^n$, $y_i \in R$, and $i = 1, \cdots, N$. LS-SVM technique utilizes $\sum_i^N \xi_i^2$ to substitute $\sum_i^N \xi_i$ of the standard SVM, and the optimization problem is formulated as following,

$$\begin{cases} \min \quad J(\omega, \xi) = \frac{1}{2}\omega^T \omega + \frac{1}{2}\gamma \sum_{i=1}^N \xi^2 \\ \text{s.t.} \quad y_i = \omega^T \phi(x_i) + \theta + \xi_i \end{cases} \quad (7)$$

where $\phi(\cdot)$ is a nonlinear mapping function. Using Lagrange multipliers technique and KKT complementary condition, we can transform the above optimization problem into the problem of solving the following linear equation,

$$\begin{bmatrix} \Omega + \gamma^{-1} I & l_N \\ l_N^T & 0 \end{bmatrix} \begin{bmatrix} \alpha \\ \theta \end{bmatrix} = \begin{bmatrix} y \\ 0 \end{bmatrix} \quad (8)$$

where $l_N = [1, 1, \cdots, 1]^T$, $\alpha = [\alpha_1, \alpha_2, \cdots, \alpha_N]^T$ and $y = [y_1, y_2, \cdots, y_N]^T$. Ω is a square matrix with entries $\Omega_{ij} = \psi(x_i, y_j) = \phi(x_i)^T \phi(x_i)$, $i, j = 1, \cdots, N$. $\psi(\cdot)$ is a kernel function which satisfies the Mercer's theorem. α and θ can be obtained by solving equation (8), and the linear regression model can be represented by the following formula:

$$y(x) = \sum_{i=1}^N \alpha_i \psi(x, x_i) + \theta \quad (9)$$

In this research, Gaussian kernel function is chosen as ψ, so the expression of ψ is expressed as following,

$$\psi(x, y) = \exp(\frac{-\|x - y\|^2}{2\sigma^2}) \quad (10)$$

Finally, the resulting LS-SVM model for function estimation becomes [15],

$$f(x) = \sum_{i=1}^{N} y_i \alpha_i \psi(x, x_i) \qquad (11)$$

And the final decision function becomes sgn$[f(x)]$, where

$$\text{sgn}(x) = \begin{cases} 1, x > 0 \\ -1, x < 0 \end{cases}.$$

In the application of bivariate process, $p = 2$, the class numbers of patterns are set to 4, that is, $y_i \in \{1, 2, 3, 4\}$.

LS-SVM technique can only construct pattern recognizer used for two-class datasets. As for the K-class classification problem ($K = 2^p$ in this paper), the "one against one" [16] classification algorithm is used for constructing a K-class pattern recognizer. The "decision blackout area" problem of this algorithm will be reflected in a number of indicators, e. g. , accuracy rate, missing rate, and false rate.

The parameters γ in equation (7) and σ in equation (10) have significant influence on the performance of the LS-SVM pattern recognizer. Regularization parameter γ and kernel parameter σ are tuning parameters to be optimized. For simplicity, we employ "cross-validation" method [17] to select appropriate values for the two parameters.

4.2 Method of generating training dataset

The best way to get training datasets is to collect various data from real-world manufacturing systems, but it costs too much. In this regard, simulation is an effective and useful alternative to investigate into the potential problems associated with the normal/abnormal patterns in a multivariate manufacturing system [18]. In this research, Monte-Carlo simulation is used to generate pseudo-random variates from a multivariate normal distribution whose mean is μ_0 and whose covariance matrix is $\sum$ (a $p \times p$ matrix, $p = 2$ in this paper), and it is assumed that the mean μ_0 of the process remains constant. As described in section 3.2, by manipulating the values of a and b, the desired N sets of pattern data ($X_{1k}, X_{2k}, \cdots, X_{nk}, k = 1, 2, \cdots, N$) can be simulated. In this research, a and (or) b are (is) either generated randomly with value range of $[1.5, 3]$ or set to 1. For instance, in order to generate $(0,1)$ pattern data of covariance matrix, a is set to 1, and b is a randomly generated number between 1.5 and 3. The simulation can be achieved through functions of MATLAB software. Meanwhile, it is essential that all the generalized variances of training data for abnormal patterns must exceed the control limit of $|S|$ chart, and all the generalized variances of training data for normal pattern must not exceed the control limit of $|S|$ chart. In this way, $|S|$ chart plays a role of variance shifts detector.

4.3 Input vector of LS-SVM

In this bivariate process study, appropriate input feature vectors should be used for training the LS-SVM pattern recognizer. Represent $X_{1k}, X_{2k}, \cdots, X_{nk}$ for the sample data at the kth sampling, the input feature vector V_k can be constructed through the following formula,

$$V_k = (S_{1k}^2, S_{12k}, S_{2k}^2) \tag{12}$$

Where, n is the subgroup size at the kth sampling, S_{12k} indicates the sample covariance of the two quality characteristics, S_{1k}^2 and S_{2k}^2 are the sample variances of the two quality characteristics respectively. After N times of sampling, $k = 1, 2, \cdots, N$, there are N input feature vectors can be got to train the LS-SVM pattern recognizer.

The concern of this research is the dispersion of the process, so the sample variances are selected as the elements of the input feature vectors to train the LS-SVM pattern recognizer. Meanwhile, |S| chart works as a variance shifts detector to determine whether the bivariate process is in control according to the observations $X_{1k}, X_{2k}, \cdots, X_{nk}$.

5 Simulation study and performance evaluation

In this section, a bivariate numerical case is presented to evaluate the performance of the model through simulation experiment. For simplicity and without loss of generality, it is assumed that the in-control process follow 2-dimensional combination normal distribution with mean zero, variance one and correlation coefficient ρ. In actual applications, the values of the quality characteristics can be scaled to fit the assumption using formula $\frac{X-\mu}{\sigma}$. In the process of the experiment, both the training data and the test data of the four patterns of covariance matrix were simulated in the same way as described in section 4.2. In one experiment, the subgroup size was the same to both the training data and the test data.

To get better performance, γ and σ should be optimized, and finally we get $\gamma = 1$ and $\sigma = 0.3$. With number of training subgroup N=15, subgroup size n=10 and different values of correlation coefficient ρ, the experiment was done and the experiment results are summarized in Table 1.

From Table 1, it can be seen that the performance of the LS-SVM recognizer does not change obviously in different values of ρ among x_1 and x_2, and the LS-SVM recognizer can provide satisfactory performance for the whole range of the investigated ρ's.

Note that this conclusion requires the assumption that the occurrence of assignable cause does not shift the correlation coefficient ρ among x_1 and x_2. If this assumption is not satisfied, the performance of the LS-SVM recognizer may degrade severely. It is a problem need to further study in the future.

Tab 1 Performance of the pattern recognizer
(Number of training subgroup $N=15$, Subgroup size $n=10$)

ρ	Patterns of 1000 test vector data	\multicolumn{4}{c	}{Classified situation of 1000 test vector data}	\multicolumn{3}{c	}{Classification percentage (%)}
		(0,0)	(0,1)	(1,0)	(1,1)	Accuracy rate	False rate	Missing rate
0.2	(0,1)	9	866	12	113	86.6	12.5	0.9
	(1,0)	7	1	948	44	94.8	4.5	0.7
	(1,1)	2	140	167	691	69.1	30.7	0.2
0.4	(0,1)	38	911	35	16	91.1	5.1	3.8
	(1,0)	38	0	783	179	78.3	17.9	3.8
	(1,1)	11	141	166	682	68.2	30.7	1.1
0.6	(0,1)	18	884	64	34	88.4	9.8	1.8
	(1,0)	4	2	888	106	88.8	10.8	0.4
	(1,1)	2	200	178	620	62.0	37.8	0.2
0.8	(0,1)	15	903	0	82	90.3	08.2	1.5
	(1,0)	24	7	854	115	85.4	12.2	2.4
	(1,1)	3	135	96	766	76.6	23.1	0.3

Table 1 also indicates that the recognition performance reduces in recognizing pattern (1,1) of the covariance matrix, at the same time, the LS-SVM recognizer shows similar performances on recognizing pattern (0,1) and pattern (1,0).

For comparison purposes, the simulation experiment was also done under different number of subgroup N, subgroup size n (see Figure 2 and Figure 3). The results in Figure 2 and Figure 3

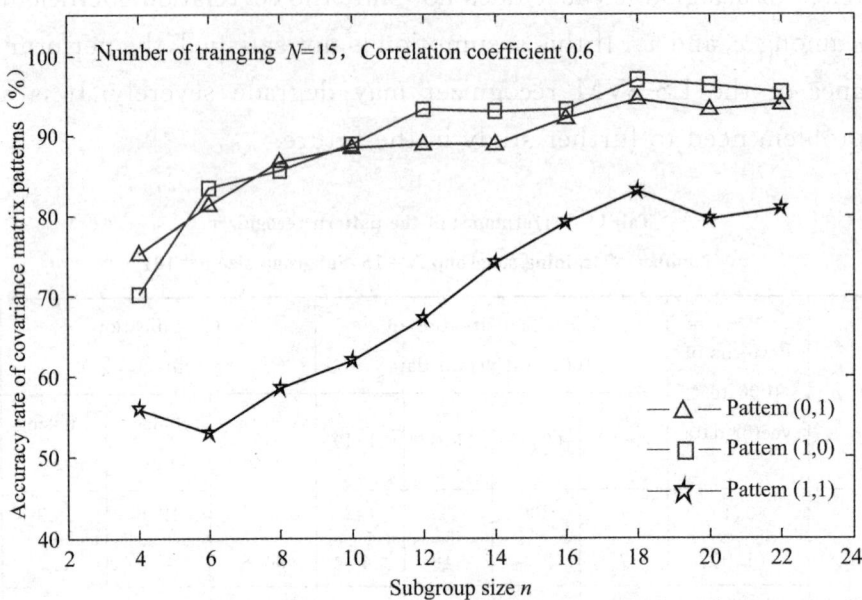

Fig 2　Relationship curve between accuracy rate-subgroup size *n*

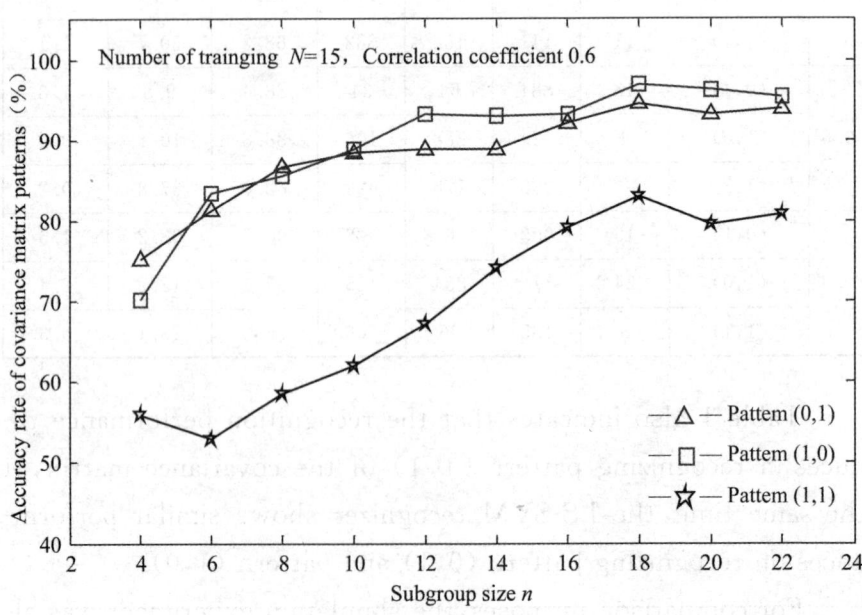

Fig 3　Relationship curve between accuracy rate-*N*（number of training subgroup）

show that higher accuracy rates can be obtained by selecting bigger N and n appropriately.

6 An example of application

This section presents a real world example, which was given by Montgomery [1] and researched by Machado et al. [7], to illustrate the usage of the proposed diagnosis model for bivariate process. In the example, the two quality characteristics to monitor are the tensile strength (psi) and the diameter of a textile fiber ($\times 10^{-2}$ inch), and the values of the mean vector and the covariance matrix are considered to be known from historical data:

$$\mu_0 = \begin{bmatrix} 115.59 \\ 1.06 \end{bmatrix}, \Sigma_0 = \begin{bmatrix} 1.23 & 0.79 \\ 0.79 & 0.83 \end{bmatrix}$$

The correlation coefficient of the two quality characteristics is $\rho = 0.782$, and it remains constant in this example. Initially the bivariate process is in control. Afterwards the assignable cause appears with out-of-control covariance matrix Σ_1,

$$\Sigma_1 = \begin{bmatrix} 3.69 & 1.368 \\ 1.368 & 0.83 \end{bmatrix}$$

Using the process parameters $n=6, p=2, \mu_0$ and Σ_0, we first calculate the control limits of $|S|$ chart and get $UCL = 1.316$ and $LCL = 0$. Then the LS-SVM recognizer can be well trained with parameters $N=15, n=6, \gamma=1$ and $\sigma=0.3$. At last, we diagnose the covariance matrix of bivariate process using the trained LS-SVM recognizer.

To illustrate the application of the proposed model, 15 in-control samples of the bivariate process are initially simulated with covariance matrix Σ_0. Soon after, the process variability abnormally changes with changed covariance matrix Σ_1, and the quality engineers get the $|S|$ chart which is shown in Figure 4.

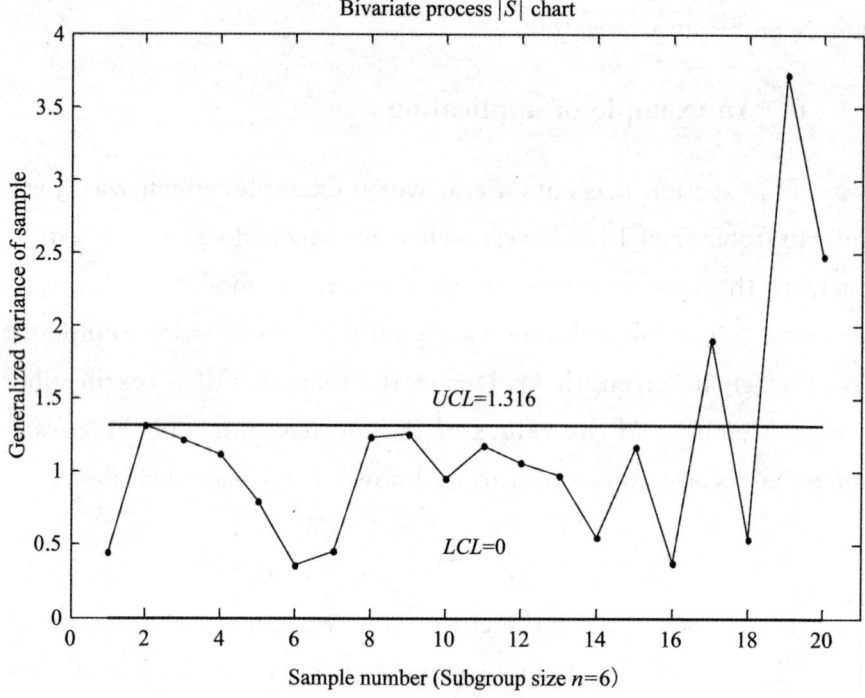

Fig 4 |S| chart of the example

From Figure 4, |S| chart detects process abnormality at samples 17,19 and 20. At these sample points, quality engineers feed the feature vectors of these samples into the LS-SVM recognizer to determine which variable(s) is (are) responsible for it. The results of pattern recognition indicate that the three out-of-control patterns are all type (1,0). This shows that the first variable is responsible for the process abnormality of this time. When speaking of the variable that is responsible for the process abnormality, the real implication is that the value of the variable is affected by assignable cause(s). Accordingly, based on the information provided by LS-SVM pattern recognizer, quality engineers search for the assignable cause(s) and take necessary corrective actions to bring the out-of-control process back to the in-control state.

This diagnosis results agree with those in research [7], where statistical method is used to diagnose the bivariate process. Compared with statistical method, LS-SVM method does not actually need the prior knowledge of the process distribution, and has a wider applicability in practice.

7　Conclusion

In multivariate process, it is crucial to identify and eliminate the assignable causes quickly. When process has a multivariate normal distribution, it is necessary to monitor not only the abnormality in process mean but also the abnormality in process dispersion. Much work has been done on monitoring the process through overall statistics, but less work was done on finding the root cause of the process abnormality.

In this article, we investigate a model consisting of $|S|$ chart and LS-SVM pattern recognizer to diagnose the dispersion changes in bivariate process. When an alarm signal is triggered by $|S|$ chart, the LS-SVM recognizer works as a supplement to provide useful information about which variable(s) is (are) responsible for the process abnormality. The performance of the model is evaluated through simulation, and the primary results of the simulation are promising. An application example is given to illustrate the usage and effectiveness of this model.

This research converts the problem of identifying the assignable cause of process abnormality into the problem of patterns recognition of covariance matrix by using LS-SVM pattern recognizer. LS-SVM techniques have an obvious advantage when it is applied to production field. Namely, compared with traditional ANN methods, LS-SVM recognizer can be trained in the case of smaller number of samples. Although the proposed model is a particular application in bivariate process, it can be potentially applied to multivariate SPC in general. In this aspect, future work may involve how to properly define the patterns of covariance matrix of multivariate process, and how to further improve the accuracy rate of patterns recognition.

Acknowledgement

This work was supported by two National Natural Science Foundation of China, Grant No. 70931002 and No. 70672088.

References

[1] Montgomery DC. Introduction to Statistical Quality Control. Wiley: New York, 2004.

[2] Hotelling H. Multivariate quality control. Techniques of Statistical Analysis, Eisenhart C, Hastay M, WallisWA (eds.). McGraw-Hill: New York, 1947.

[3] R B Crosier. Multivariate generalizations of cumulative sum quality control schemes [J], Technometrics, 1988, 30: 291 −303.

[4] Lowry CA, Woodall WH, Champ CW, Rigdon SE. A multivariate exponentially weighted moving average control chart [J]. Technometrics, 1992, 34: 46−53.

[5] Lowry CA, Montgomery DC. A review of multivariate control charts [J], IIE Transactions, 1995, 27: 800−810.

[6] Alt F B. Multivariate quality control. Encyclopedia of Statistical Sciences, Kotz S, Johnson NL (eds.). Wiley: New York, 1985.

[7] M A G Machado, A F B Costa, M A Rahim. The Synthetic Control Chart Based on Two Sample Variances for Monito-

ring the Covariance Matrix [J], Qual. Reliab. Engng. Int. ,2008, 25(5):595—605.

[8] Chinyao Low, Chih-Ming Hsu, Fong-Jung Yu. Analysis of variations in a multi-variate process using neural networks [J], Int J Adv Manuf Technol, 2003, 22:911—921.

[9] Jian-bo Yu, Li-feng Xi. A neural network ensemble-based model for on-line monitoring and diagnosis of out-of-control signals in multivariate manufacturing processes [J], Expert Systems with Applications, 2009, 36:909—921.

[10] Ruey-Shiang Guh. On-line Identification and Quantification of Mean Shifts in Bivariate Processes using a Neural Network-based Approach [J], Qual. Reliab. Engng. Int. ,2007, 23:367—385.

[11] Nello Cristianini, John Shawe-Taylor. An Introduction to Support Vector Machines and Other Kernel-based Learning Methods [M]. Cambridge University Press, 2005.

[12] J A K Suykens, J Vandewalle. Least Squares Support Vector Machine Classifiers [J], Neural Processing Letters, 1999, 9:293—300.

[13] S. Bersimis, S. Psarakis, J. Panaretos. Multivariate Statistical Process Control Charts: An Overview [J]. QUALITY AND RELIABILITY ENGINEERING INTERNATIONAL, 2007, 23:517—543.

[14] F. Zorriassatine, A. Al-Habaibeh, R. M. Parkin, M. R. Jackson, J. Coy. Novelty detection for practical pattern recognition in condition monitoring of multivariate processes: a case study [J]. Int J Adv Manuf Technol, 2005, 25:954—963.

[15] C. L. Huang, C. J. Wang. A GA-based feature selection and parameters optimization for support vector machines [J]. Expert Systems with Applications, 2006, 31:231—240.

[16] Bo Liu, Zhifeng Hao, Xiaowei Yang. Nesting algorithm for multi-classification problems [J]. Soft Computing, 2007, 11: 383-389.

[17] Senjian An, Wanquan Liu, Venkatesh, S. Fast cross-validation algorithms for least squares support vector machine and kernel ridge regression [J]. Pattern Recognition, 2007, 40(8): 2154-2162.

[18] F. Zorriassatine, J. D. T. Tannock, C. O. Brien. Using novelty detection to identify abnormalities caused by mean shifts in bivariate processes. Computers & Industrial Engineering, 2003, 44: 385-408.

An intelligent method of change-point detection based on LS-SVM algorithm

Cheng Zhiqiang

School of Management and Economics, North China University of Water Resources and Electric Power, Henan Zhengzhou, 450011, P. R. China.

czqme@163.com

Abstract: The change-point detection problem is an important consideration in many application areas. This paper aims to address this issue in autocorrelated process, where traditional statistical methods are not applicable. Based on LS-SVM pattern recognizer, this paper develops an intelligent method for solving the problem of change-point detection, and the proposed model is applied to detect change-point of process mean-shift in auto-correlated time series process. In this research, LS-SVM algorithm and moving window method are used to detect the location of the mean shift signal. The LS-SVM pattern recognizer is designed and the performance of the recognizer is evaluated in terms of Accuracy Rate. Results of simulation experiment show that the proposed intelligent model is an effective method to detect change-point in the mean of ARMA data series. Compared with the traditional statistical methods for change-points detection, the proposed intelligent method can work well without the need for the prior information of the process data. Though the proposed method in this paper is only used in solving the problem of single change-point detection in the mean, it can be potentially extended and used in solving the practical problems of multiple change-point detection.

Key words: change-points detection; least squares support vector machine (LS-SVM); moving window; pattern recognition; autoregressive moving average (ARMA) process

1 Introduction

The change-point detection problem deals with a stochastic process, where the process distribution abruptly changes at some unknown instants. The detection of change-points is an important consideration in many application areas. As far as the quality management area is concerned, when the process parameters (e. g. , mean, variance) of the process are suspected to have an abrupt shift, the detecting problems are similar to the sequential change-point detection [1]. In ecology, change-point detection problem is instead referred to as "boundary or edge detection" problem to identify the changes take place along the line transect [2,3]. In finance economics, Change-point detection is applied to predict the interest rates [4]. In addition, a change-point problem includes the abrupt change in adaptive signal processing, the fault detection in technological processes, the incidence of a disease in epidemiology, the temperature change in meteorology, and so on.

The research of the change-points problem dates back to Page [5] which identified the existence of single change point. Page's CUSUM (cumulative sums) chart is a familiar technology for detecting change-points in SPC (Statistical Process Control) methodology. Lai and others [6] considered the change-point problem in the more general situation of dependent random variables. Most of the applications are interested in the cases of composite hypotheses for the problem of sequential change-point detection. Usually, the well-studied change-point model is characterized by the probabilistic structure of the monitored process. So far, most of the change point problems have been discussed statistically in the literature [7,8,9,10,11,12,13]. The common form of the statistical approach can be found in Lai [14]. Statistical techniques to address change-point problem require a priori knowledge (e. g. , parameters and distribution of the process) of the process. The prerequisite of a priori knowledge about the process usually cannot be easily satisfied or validated, and this makes it difficult to apply statistical techniques in a particular application. Also, when the process sequential data are autoregressive, to the best of my knowledge, most of the statistical methods for change point detection are no longer available.

This study deviates from other previous studies by using artificial intelligence (AI) techniques as an alternative to detect the change point in autoregressive process. AI methods do not rely on the a priori knowledge of the process data and have a wider range of applicability, and this is the main advantage of the intelligent method to address the change-point problem in this paper. The representative intelligent techniques mainly include Artificial Neural Networks (ANN) [15,16] and Support Vector Machine (SVM) [17]. ANN techniques require large amounts of training sample data when they are used for process monitoring. At the same time, almost all the manufacturing enterprises are unwilling to afford excessive cost for sample data anyway. ANN is not always feasible when it is used in a real application. It is necessary for quality practitioners to study the process monitoring techniques which can use as less samples as possible. Support Vector Machine is quite an efficient technique developed in statistical learning theory, its effective application can be found in the area of solving problems of small training samples, nonlinear and high dimension. Furthermore, least squares support vector machine (LS-SVM) [18,19] is an extension of standard SVM with faster computational speed. SVM techniques can be used in pattern recognition with small sample size. Using LS-SVM pattern recognizer, this research formulates the problem of locating the position of the change-point in process mean into the problem of pattern recognition. The performance of the LS-SVM pattern recognizer is evaluated based on Accuracy Rate (AR).

The organization of this paper is as follows. Section 2 briefly introduces the process model investigated in this paper. Section 3 presents a new intelligent model of detecting the change-point. Section 4 briefly describes LS-SVM techniques and its working principle. The performance of the proposed model is evaluated in Section 5. Section 6 gives some concluding remarks.

2 Description of process model

The interest of this research lies in detecting the single change-point in the mean of a process. Mean shift is the most common abnormal change of a process. For a practical process, process data usually show autocorrelation. In this situation, an autoregressive moving average (ARMA) model is often used to represent the autocorrelation of the data. Thus, this article uses LS-SVM pattern recognizer to detect the change-point in the mean of an ARMA process.

This article studies an open-loop time-series ARMA model, which is shown in Figure 1. This model well signifies the situation of single working procedure.

The model in Figure 1 includes two parts: the dynamics process and the disturbance noise process. $G(B)$ is the transfer function of the dynamics process, the input of $G(B)$ is μ_t, the output of $G(B)$ is y_t. When disturbance n_t does not exist, the target value of the steady-state output Y_t of the dynamics process is set to T, without of generalization, let $T = 0$ in this article. Disturbance n_t consists of mean shift signal and the colored noise $\frac{\theta(B)}{\phi(B)}a_t$, which is represented by $ARMA(p,q)$ (Auto Regressive Moving Average) model. Namely, $n_t = \frac{\theta(B)}{\phi(B)}a_t + \gamma f(t)$. B is the backshift operator, $B^K X_t = X_{t-K}$. $\phi(B)$ and $\theta(B)$ are referred to as the AR and MA polynomial, and are parameterized as $\phi(B) = 1 - \phi_1 B - \phi_2 B^2 - \cdots - \phi_p B^p$ and $\theta(B) = 1 - \theta_1 B - \theta_2 B^2 - \cdots - \theta_q B^q$ respectively. $f(t)$ indicates the nature of the mean shift signal, γ is the amplitude of the mean shift signal.

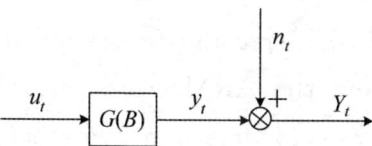

Fig 1　The structure of the ARMA model

When the dynamics process is in steady state and the mean shift has not yet appeared,

$$Y_t = T + \frac{\theta(B)}{\phi(B)}a_{t0}. \qquad (1)$$

Here, the zero in a_t is added to indicate that the a_{t0} values are just residuals.

When the dynamics process is in steady state and the mean shift has appeared at some unknown time τ,

$$Y_t = T + \frac{\theta(B)}{\phi(B)}a_t + \gamma f(t), t \geqslant \tau. \qquad (2)$$

3 Intelligent model of detecting change-point

Based on Figure1, this article proposes an intelligent model to detecting the change-point in the mean of an ARMA process in Figure 2. Figure 2 shows a schematic for detecting the single change-point in the mean of a ARMA process using LS-SVM pattern recognizer.

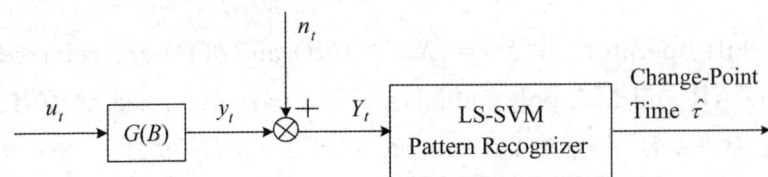

Fig 2 Intelligent model for detecting the change-point in the mean

In Figure 2, we assume that the starting time of the process is $t=0$, and the LS-SVM recognizer is applied to analyze the observational data from the ARMA process. The function of the LS-SVM recognizer is to determine whether the process mean changes abruptly and to further indicate the location of the change-point. If no change-point exists in the process mean, the LS-SVM recognizer indicate that the process is in a normal state; if change-point exists in the process mean, the LS-SVM recognizer indicate that the process is in a abnormal state, and estimate the location of the change-point (i. e. , the occurrence time of the change-point). When LS-SVM recognizer first issues a signal of abnormal pattern which signify the occurrence of change-point in the process mean at a certain time, which is denoted as $\hat{\tau}$, then time $\hat{\tau}$ is considered to be the 'likely' time of the change in the process mean. The essentiality of change-point analysis method is classifying the time series. Namely, the total time series is piecewise analyzed, and the occurrence of change-point is judged according to the property of each segment of time series. Please note in Figure 2 that we identify the occurrence time of the mean shift signal based on LS-SVM techniques in ARMA processes, where traditional statistical method (e. g. , CUMUM chart) is not suitable to be used to detect change point. Though the problem of identifying the occurrence time of the mean shift signal in our study is similar to the problem of change point detection, statistical methodology which need to know the precise distribution of the processes is not suitable for use in correlated processes in our ARMA model.

To identify the time-of-occurrence of the change-point in ARMA process mean, moving windows method is used to deal with the observational data. With the running of the ARMA process, a series of observational data $Y_1, Y_2, \cdots, Y_n, \cdots$ can be obtained. By designing suitable moving window, with window length (which is also called window size) l_w and moving step interval h_m, we can get the following new time series,

$$\begin{cases} Y'_1 = \{Y_1, Y_2, \cdots, Y_{l_w}\} \\ Y'_2 = \{Y_{1+h_m}, Y_{2+h_m}, \cdots, Y_{l_w+h_m}\} \\ \vdots \\ Y'_d = \{Y_{1+(d-1)h_m}, Y_{2+(d-1)h_m}, \cdots, Y_{l_w+(d-1)h_m}\} \\ \vdots \end{cases},$$

$$d = 1, 2, 3, \cdots \qquad (3)$$

where, d is the moving number of the moving window. The normal pattern and the abnormal pattern of time series Y'_d should be first investigated and defined respectively. In the definitions of the two patterns, the time series data Y'_d of the normal pattern only contain the data of the normal process (i. e. , no change-point exists in process mean), to be specific, the time series data Y'_d of the normal pattern can be generated according to Equation

(1) by simulation. The time series data Y'_d of the abnormal pattern contain the data information of the mean shift signal, to be specific, the time series data Y'_d of the abnormal pattern can be generated according to Equation (2) by simulation.

By classifying the patterns of the time series Y'_d and analyzing the change of the patterns, we can get the information about the occurrence time of the change-point signal. When the pattern of time series Y'_d changes from the normal pattern to the abnormal pattern, it can be determined that the change-point signal occur in the moving window of the abnormal pattern.

4 LS-SVM pattern recognizer

4.1 LS-SVM principle

Let x_i be a set of data points and y_i indicate the corresponding classes, where $x_i \in R^n$, $y_i \in R$, and $i=1,\cdots,N$. LS-SVM technique utilizes $\sum_{i}^{N}\xi_i^2$ to substitute $\sum_{i}^{N}\xi_i$ of the standard SVM, and the optimization problem is formulated as follows,

$$\begin{cases} \min \quad J(\omega,\xi) = \frac{1}{2}\omega^T\omega + \frac{1}{2}C\sum_{i=1}^{N}\xi_i^2 \\ \text{s. t.} \quad y_i = \omega^T\phi(x_i) + \theta + \xi_i \end{cases} \quad (4)$$

where $\phi(\cdot)$ is a nonlinear mapping function. Using Lagrange multipliers technique and KKT complementary condition, we can transform the above optimization problem into the problem of solving the following linear equation,

$$\begin{bmatrix} \Omega + C^{-1}I & l_N \\ l_N^T & 0 \end{bmatrix} \begin{bmatrix} \alpha \\ \theta \end{bmatrix} = \begin{bmatrix} y \\ 0 \end{bmatrix}, \quad (5)$$

where $l_N = [1,1,\cdots,1]^T$, $\alpha = [\alpha_1,\alpha_2,\cdots,\alpha_N]^T$ and $y = [y_1,y_2,\cdots,y_N]^T$. Ω is a square matrix with entries $\Omega_{ij} = \psi(x_i,y_j) =$

$\phi(x_i)^T \phi(x_j), i,j = 1,\cdots,N$. $\psi(\cdot)$ is a kernel function which satisfies the Mercer's theorem. α and θ can be obtained by solving equation (5), and the linear regression model can be represented by the following formula:

$$y(x) = \sum_{i=1}^{N} \alpha_i \psi(x, x_i) + \theta. \qquad (6)$$

In this research, Gaussian kernel function is chosen as ψ, so the expression of ψ is the following,

$$\psi(x, y) = \exp\left(\frac{-\|x-y\|^2}{2\sigma^2}\right). \qquad (7)$$

Finally, the resulting LS-SVM model for function estimation becomes [20],

$$f(x) = \sum_{i=1}^{N} y_i \alpha_i \psi(x, x_i). \qquad (8)$$

The final decision function becomes $\mathrm{sgn}[f(x)]$, where $\mathrm{sgn}(x) = \begin{cases} 1, x > 0 \\ -1, x < 0 \end{cases}$.

LS-SVM technique can only construct pattern recognizer used for two-class datasets. If a study involves K-class classification problem, the "one against one" [21] classification algorithm is used for constructing a K-class pattern recognizer. The "decision blackout area" problem of this algorithm will be evaluated by some indicators, e. g., accuracy rate, missing rate, and false rate.

The parameters C and σ have significant influence on the performance of the LS-SVM pattern recognizer. For simplicity, we employ "cross-validation" method [22] to select appropriate values for the two parameters.

4.2 The training of LS-SVM recognizer

The training of LS-SVM pattern recognizer is an important

task that needs to be addressed before its implementation. The best way to get training datasets is collecting various data from real production systems, but it costs too much. In this regard, Monte-Carlo simulation is an effective alternative to generate the required training data and test data. In our simulation, the data series of the normal pattern can be generated according to Equation (1), the data series of the shifted pattern can be generated according to Equation (2). The simulation is preformed under different experimental condition, such as different values of γ and different window length l_w, etc. In particular, for a signal with type of mean shift, the mid point of the data series of the shifted pattern is designed to be the shift point. Mean shift signal is one of the most common abnormal signals in the manufacturing processes, and is usually expressed as

$$f(t) = \begin{cases} 1, t \geq \tau \\ 0, t < \tau \end{cases}. \tag{9}$$

In this case, parameter γ signifies the magnitude of the mean shift signal. In order to visually reveal the normal/abnormal pattern data in our simulation, Figure 3 is drawn to show the shape of the normal/abnormal pattern data under the experimental condition of window size $l_w = 28$, magnitude of mean shift $\gamma = 2.5$, and $ARMA(1,1)$ parameters $\phi_1 = 0.6, \theta_1 = -0.4$.

5 Performance evaluation of the model

Considering the model in Figure 2, this research takes $ARMA(1,1)$ model as an example to evaluate the performance of the proposed method. As for the $ARMA(1,1)$ model, it involves two parameters, namely, ϕ_1 and θ_1. The researched abnormal signal is presumed to be mean shift signal, in this sense, the abnormal pattern can be also called shifted pattern.

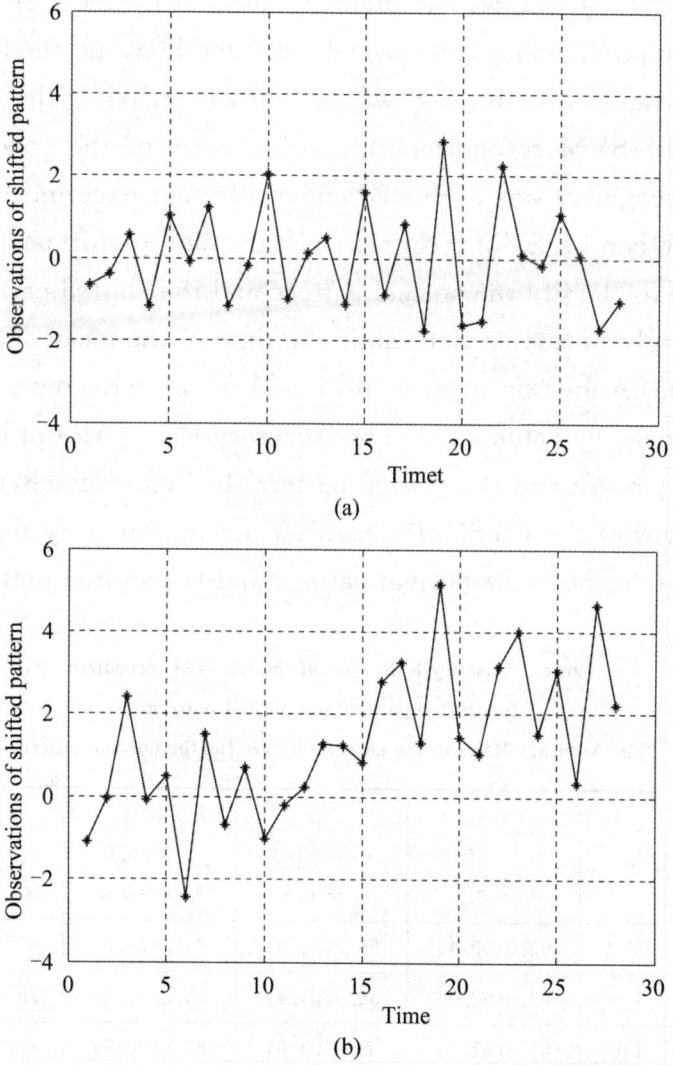

Fig 3 (a) Training data of normal pattern; (b) Training data of shifted pattern

In this research, to train the LS-SVM pattern recognizer under different parameters setting, 20 training pattern data were respectively simulated for the normal pattern and the abnormal pattern (in other words, shifted pattern). Correspondingly, another 1000 test pattern data were respectively simulated for the two patterns on the same experimental parameters. The parameters include γ, l_w, ϕ_1 and θ_1. Parameters C and σ of LS-SVM recognizer were optimized using "cross validation" method, and the final optimized values are obtained with $C=10, \sigma=1$. After the training of the LS-SVM recognizer, the performance of the trained LS-SVM recognizer was evaluated under different experimental condition. When LS-SVM recognizer obtains high performance, it can be used to classify the normal pattern and the shifted pattern and accordingly to further determine the time of the mean shift.

The evaluation results of LS-SVM pattern recognizer is summarized in Table 1 based on Accuracy Rate (AR) of both the normal pattern and the shifted pattern. In this research, the performances of the LS-SVM recognizer are indicated by the Accuracy Rate of both the normal pattern and the shifted pattern.

Tab 1 Accuracy Rate (%) of the LS-SVM recognizer for shifted pattern and normal pattern

(The Accuracy Rates in the parentheses are for the normal pattern)

γ	l_w	(a) $\phi_1=-0.6$ $\theta_1=0.4$	(b) $\phi_1=-0.2$ $\theta_1=0.8$	(c) $\phi_1=0.6$ $\theta_1=-0.4$	(d) $\phi_1=0.2$ $\theta_1=-0.8$
1	6	62.3(59.4)	56.2(52.3)	76.3(73.0)	67.1(75.9)
	10	62.9(62.2)	56.5(51.4)	80.3(73.3)	79.6(70.2)
	14	61.5(65.7)	58.6(53.5)	80.6(80.9)	75.7(71.9)
	18	60.6(63.7)	60.2(52.9)	77.9(78.9)	75.6(67.6)
	22	69.3(56.5)	56.7(46.8)	81.3(65.1)	79.0(60.6)
	26	65.2(50.8)	63.7(37.4)	75.3(72.2)	89.3(35.9)
	30	86.2(30.5)	85.4(19.2)	95.5(32.1)	96.5(27.0)
2	2	84.4(75.0)	74.4(71.2)	74.7(83.4)	75.5(74.3)
	4	75.8(69.2)	70.1(68.9)	91.6(94.0)	96.8(90.2)
	6	76.9(71.8)	78.5(66.9)	95.6(94.9)	93.9(97.7)

续表

γ	l_w	(a) $\phi_1=-0.6$ $\theta_1=0.4$	(b) $\phi_1=-0.2$ $\theta_1=0.8$	(c) $\phi_1=0.6$ $\theta_1=-0.4$	(d) $\phi_1=0.2$ $\theta_1=-0.8$
	8	74.8(71.1)	65.6(77.0)	99.0(98.8)	96.3(95.3)
	10	80.5(63.5)	63.5(69.4)	99.5(99.1)	98.5(96.3)
	12	75.8(79.6)	65.0(47.9)	98.9(99.5)	98.2(97.1)
	14	80.8(75.0)	69.5(63.7)	99.6(99.4)	99(97.8)
	16	78.9(84.9)	64.8(59.9)	99.4(98.9)	96.8(96.1)
	18	79.5(80.2)	77.2(55.9)	99.2(97.2)	98.2(93.2)
	20	86.4(79.1)	83.8(49.2)	100(94.9)	99.3(87.9)
	22	87.3(69.4)	82.4(36.0)	99.9(88.1)	99.2(75.9)
	24	83.5(73.6)	79.1(51.9)	100(80.6)	99.9(58.5)
	26	87.3(68.9)	68.7(41.1)	100(62.2)	100(51.2)
	28	97.8(39.3)	81.1(39.0)	100(57.3)	100(37.4)
	30	96.6(41.2)	89.2(28.4)	100(44.3)	100(20.1)
3	6	88.5(84.0)	84.2(82.3)	100(99.8)	100(99.9)
	10	91.8(81.8)	81.0(81.7)	100(99.9)	99.9(98.6)
	14	89.4(93.1)	80.3(84.8)	100(99.9)	99.8(98.0)
	18	93.0(88.2)	87.0(77.8)	100(98.1)	99.8(93.2)
	22	96.3(86.1)	89.6(70.4)	100(91.2)	100(79.0)
	26	98.7(64.7)	91.2(53.8)	100(66.8)	100(55.2)
	30	99.9(45.4)	96.0(21.1)	100(50.1)	100(29.6)

From the experiment results of Accuracy Rates in Table 1, some conclusions can be drawn as following:

1) With the increase of the magnitude of the mean shift, the performances of the LS-SVM recognizer get better. This means the proposed method is more suited to detect bigger mean shift signal. In this research, when the magnitude value of the mean

shift is larger than 2, the LS-SVM recognizer have good performance of Accuracy Rate.

2) The simulation results listed in Table 1 show that the Accuracy Rates (of both the normal pattern and the shifted pattern) vary over a range depending upon the ARMA parameters ϕ_1 and θ_1. This is because the variance of the time series data of $ARMA(1,1)$ model vary with the changes of the values of the parameters ϕ_1 and θ_1.

3) The LS-SVM recognizer is only effective if the Accuracy Rates of both the shifted pattern and the normal pattern are high enough. To obtain high Accuracy Rates of both the shifted pattern and the normal pattern, the window size l_w of the moving window should be selected rationally. In this research, simulation experimental method is used as an effective way to select proper window size. To illustrate this method, for the experiment results and the values of parameters ϕ_1 and θ_1 in case (f) (when $\gamma=2$) of Table 1, the relationships of Accurate Rate and window size are drawn in Figure 4. From Figure 4, it is can be seen obviously that the Accuracy Rates of both the shifted pattern and the normal pattern are very high when the values of window size are 6, 8, 10, 12, 14, 16, or 18. In this situation, these values of window size are appropriate choices to get good performance of LS-SVM pattern recognizer. Among these optional values of window size, the selection of small window size is clearly helpful to reduce the numbers of sampling. In addition, when the window size is excessively increased, it can be observed clearly from Figure 4 that the Accuracy Rate of the abnormal pattern nearly increases to 100 percent; meanwhile, the Accuracy Rate of the normal pattern nearly decreases below 40 percent.

Please note, the optimized parameters $C = 10$, $\sigma = 1$ of LS-

SVM pattern recognizer, which is simply obtained by cross-validation method, are applied to all the experimental conditions in Table 1. In other words, all the experiments in Table 1 commonly use the same optimized parameters $C=10, \sigma=1$ of LS-SVM pattern recognizer. It is no need to design different optimized parameters C and σ of LS-SVM pattern recognizer for different parameters ϕ_1 and θ_1 of $ARMA(1,1)$ model.

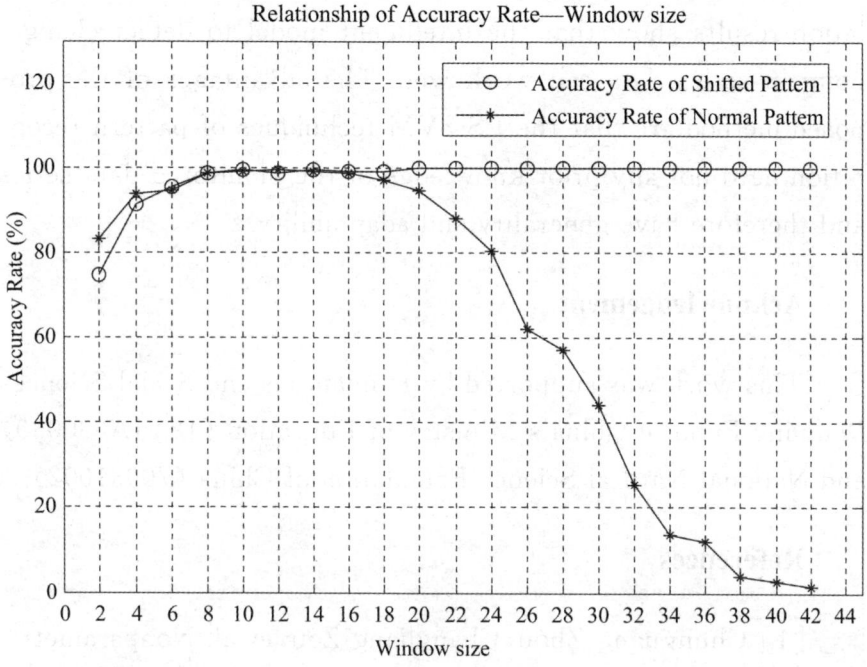

Fig 4　Relationship curve between Accuracy Rate and Window size

6　Conclusion

Lots of applications of change-points detection can be found in different areas of applied statistics. Classical statistical methods for change-points detection are interested in control the overall significance level of the monitored data, and essence of them is the statistical problem of hypothesis testing. Statistical methods for change-points detection need to know the priori informations of the monitored data, and it is very hard to satisfy on this point in many practical applications.

Based on the LS-SVM techniques of pattern recognition, this research develops an intelligent model used for change-point detection in $ARMA(1,1)$ process data. Details on designing the LS-SVM pattern recognizer have been provided. The performance of the proposed method is evaluated based on Accuracy Rate. Simulation results show that the intelligent model to detect change-point in series data can work well. The advantages of the proposed method are that the LS-SVM techniques of pattern recognition need not any prior knowledge of the monitored data series and therefore have generality and adaptability.

Acknowledgement

This work was supported by Humanities and Social Sciences Planning Fund of China's Ministry of Education (12YJA630030) and National Natural Science Foundation of China (70931002).

References

[1] Chunguang Zhou, Changliang Zou, et al. Nonparametric control chart based on change-point model. Statistical Papers, 2009, Vol. 50(1), pp13−28.

[2] I. López, M. Gámez, J. Garay, et al. Application of Change-Point Problem to the Detection of Plant Patches. Acta Biotheoretica, 2010, Vol. 58(1), pp51−63.

[3] I. López, T. Standovár, J. Garay, Z. Varga and M. Gámez (2012), Statistical detection of spatial plant patterns under the effect of forest use. International Journal of Biomathematics. DOI No:10.1142/S1793524512500544 (Forthcoming issue).

[4] Kyong Jo Oh, Ingoo Han. Using change-point detection to support artificial neural networks for interest rates forecasting. Expert Systems with Applications, 2000, Vol. 19, pp105−115.

[5] E. S. Page. Continuous inspection schemes. Biometrika, 1954, Vol. 41, pp100−116.

[6] T. L. Lai, J. Z. Shan. Efficient recursive algorithms for detection of abrupt changes in signals and systems. IEEE Trans Automat Contr, 1999, Vol. 44, pp952−966.

[7] D. M. Hawkins, P. Qiu, C. W. Kang. The Changepoint Model for Statistical Process Control. Journal of Quality Technology, 2003, Vol. 35, pp355−365.

[8] K. D. Zamba, D. M. Hawkins. A Multivariate Change Point Model for Statistical Process Control. Technometrics, 2006, Vol. 48(4), pp539−549.

[9] K. D. Zamba. A Multivariate Change-Point Model for Change in Mean Vector and/or Covariance Structure. Journal of Quality Technology, 2009, Vol. 41(3), pp285−303.

[10] Y. Ninomiya. Information criterion for Gaussian change-point model. Statist. Probab. Lett., 2005, Vol. 72, pp237−247.

[11] Jianmin Pan, Jiahua Chen. Application of modified information criterion to multiple change point problems. Journal of Multivariate Analysis, 2006, Vol. 97, pp2221−2241.

[12] Gregory Gurevich, Albert Vexler. Change point problems in the model of logistic regression. Journal of Statistical Planning and Inference, 2005, Vol. 131, pp313−331.

[13] Boris Brodsky, Boris Darkhovsky. Sequential change-point detection for mixing random sequences under composite hypotheses. Stat Infer Stoch Process, 2008, Vol. 11, pp35−54.

[14] Lai TL. Sequential analysis: some classical problems and new challenges. Stat Sin,2001,Vol. 11(2),pp303—408.

[15] R. S. Guh. On-line Identification and Quantification of Mean Shifts in Bivariate Processes Using a Neural Network-based Approach. Quality and Reliability Engineering International,2007,Vol. 23,pp367—385.

[16] Jian-bo Yu,Li-feng Xi. A neural network ensemble-based model for on-line monitoring and diagnosis of out-of-control signals in multivariate manufacturing processes. Expert Systems with Applications,2009,Vol. 36,pp909—921.

[17] Nello Cristianini,John Shawe-Taylor. An Introduction to Support Vector Machines and Other Kernel-based Learning Methods. Cambridge University Press,2005.

[18] J A K Suykens,J Vandewalle. Least Squares Support Vector Machine Classifiers. Neural Processing Letters,1999,Vol. 9,pp293—300.

[19] Suykens Johan A K,Gestel Tony Van,Brabanter Jos De. Least Squares Support Vector Machines. World Scientific Publishers,Singapore,2002.

[20] C. L. Huang,C. J. Wang. A GA-based feature selection and parameters optimization for support vector machines. Expert Systems with Applications,2006,Vol. 31,pp231—240.

[21] Bo Liu,Zhifeng Hao,Xiaowei Yang. Nesting algorithm for multi-classification problems. Soft Computing,2007,Vol. 11,pp 383—389.

[22] Senjian An,Wanquan Liu,Venkatesh,S. Fast cross-validation algorithms for least squares support vector machine and kernel ridge regression. Pattern Recognition,2007,Vol. 40(8),pp2154—62.

参考文献

[1]马义中.减小和控制多元质量特性波动的理论和方法[D].西北工业大学,2002.

[2]张根保.现代质量工程[M].北京:机械工业出版社,2007.

[3]马林,何桢.六西格玛管理[M].北京:中国人民大学出版社,2007.

[4]张公绪,孙静.新编质量管理学[M].北京:高等教育出版社,2007.

[5]G. Taguchi. Introduction to Quality Engineering [M]. Tokyo:Asian Production Organization,1986.

[6]M. S. Phadke. Quality Engineering Using Robust Design [M]. New Jersey:Prentice-Hall,1989.

[7]卓德保,徐济超.质量诊断技术及其应用综述[J].系统工程学报,2008(23):338—346.

[8]GB/T 19000—ISO 9000 国家标准.中国国家标准化管理委员会,2000.

[9]D. C. Montgomery. Introduction to Statistical Quality Control [M]. John & Sons Inc. ,1985.

[10]A. Hald. Statistical Theory with Engineering Applications [M]. John & Sons Inc. ,1952.

[11]王仲生.智能故障诊断与容错控制[M].西安:西北工业大学出版社,2005.

[12]董华.基于质量信息集成的"全质量"管理系统实现技术研究[D].合肥工业大学,2006.

[13] 鄂加强. 智能故障诊断及其应用[M]. 长沙:湖南大学出版社,2006.

[14] 王道平,张义忠. 故障智能诊断系统的理论与方法[M]. 北京:冶金工业出版社,2001.

[15] 翟敬梅,徐晓,尹春芳,谢存禧. 生产质量在线监测、诊断和控制的粗糙集模型[J]. 华南理工大学学报,2009,37(8):1—7.

[16] 何曙光,齐二石,何桢. 基于投影变换的多元统计过程控制与诊断模型[J]. 天津大学学报,2008,41(12):1512—1517.

[17] 殷建军,余忠华,李兴林,王兆卫. 基于 Kalman 滤波的过程调节与质量监控方法[J]. 浙江大学学报,2008,42(8):1419—1422.

[18] 田学民,曹玉苹. 统计过程控制的研究现状及展望[J]. 中国石油大学学报,2008,32(4):175—180.

[19] 翟敬梅,蒋梁中,谢存禧,徐晓. 生产过程质量智能化诊断方法的研究[J]. 机械科学与技术,2003,22(5):821—823.

[20] E. A. Garcia, P. M. Frank. On the relationship approaches to fault between observer parameter identification based detection[C]. Proc. of IFAC World Congress,1996:25—29.

[21] 周东华,叶银忠. 现代故障诊断与容错控制[M]. 北京:清华大学出版社,2000.

[22] 王海宇. 过程质量控制的性能评价与改进方法研究[D]. 西北工业大学,2006.

[23] D. W. Apley, Y. A. Ding. Characterization of diagnosability conditions for variances components analysis in assembly operations [J]. IEEE Transactions on Robotics and Automation,2005,2:101—110.

[24] Y. A. Ding, D. W. Apley. Guidelines for Placing Additional Sensors to Improve Variation Diagnosis in assembly Processes [J]. International Journal of Production Research,2007,45:5485—5507.

[25] G. C. Runger, R. R. Barton, Del Castillo E., W. H. Woodall. Optimal monitoring of multivariate data for fault patterns [J]. Journal of Quality Technology, 2007, 39: 159—172.

[26] D. W. Apley, H. Y. Lee. Simultaneous identification of premodeled and unmodeled variation patterns [J]. Journal of Quality Technology, 2010, 42: 36—51.

[27] Khaled Assaleh, Yousef Al-assaf. Features extraction and analysis for classifying causable patterns in control charts [J]. Computers & Industrial Engineering, 2005, 49: 168—181.

[28] C. H. Wang, W. Kuo. Identification of control chart patterns using wavelet filtering and robust fuzzy clustering [J]. Journal of Intelligent Manufacturing, 2007, 18: 343—350.

[29] 俞青峰,徐卫玉,韩兵. AR 模型在质量控制中的应用 [J]. 计算机应用与软件, 2005, 22(2): 39—41.

[30] R. S. Guh, Y. C. Hsieh. A neural network based model for abnormal pattern recognition of control charts [J]. Computers & Industrial Engineering, 1999, 36(1): 97—108.

[31] 葛哲学, 孙志强. 神经网络理论与 MATLAB R2007 实现 [M]. 北京: 电子工业出版社, 2007.

[32] D. F. Cook, C. C. Chiu. Using radial basis function neural networks to recognize shifts in correlated manufacturing process parameters [J]. IIE Transactions, 1998, 0: 227—234.

[33] A. M. Barghash, S. Nadeer, Santarisi. Pattern recognition of control charts using artificial neural networks-analyzing the effect of the training parameters [J]. Journal of Intelligent Manufacturing, 2004, 15: 635—644.

[34] T. T. El-Midany, M. A. El-Baz, M. S. Abd-Elwahed. A proposed framework for control chart pattern recognition in multivariate process using artificial neural networks [J]. Expert Systems with Applications, 2010, 37: 1035—1042.

[35] R. S. Guh, Y. R. Shiue. On-line identification of control chart patterns using self-organizing approaches [J]. International Journal of Production Researches, 2005, 43(6): 1225-1254.

[36] 昝涛, 费仁元, 王民. 基于神经网络的控制图异常模式识别研究[J]. 北京工业大学学报, 2006, 32(8): 673-676.

[37] 陈平, 槐春晶, 罗晶. 改进的 BP 算法用于控制图模式识别[J]. 机械与电子, 2005, 3: 42-44.

[38] Susanta Kumar. A study on the various features for effective control chart pattern recognition [J]. International Journal of Advanced Manufacturing Technology, 2007, 34: 385-398.

[39] 陈立强, 阎植林, 高祥有. 质量诊断的模糊专家系统[J]. 计算机仿真, 1997, 14(3): 18-24.

[40] 巩敦卫, 许世范. 基于模糊模式识别的选煤厂工序失控原因诊断系统[J]. 系统工程理论与实践, 2000, 11: 135-138.

[41] M. H. Fazel Zarandi, A. Alaeddini, I. B. Turksen. A hybrid fuzzy adaptive sampling—Run rules for Shewhart control charts[J]. Information Sciences, 2008, 178: 1152-1170.

[42] V. Vapnik. Statistical Learning Theory [M]. New York: Wiley, 1998.

[43] Nello Cristianini, John Shawe-Taylor. An Introduction to Support Vector Machines and Other Kernel-based Learning Methods [M]. London: Cambridge University Press, 2005.

[44] 李国正, 等. 支持向量机导论[M]. 北京: 电子工业出版社, 2004.

[45] 边祺, 张学工, 等. 模式识别[M]. 北京: 清华大学出版社, 2004.

[46] 杨志民, 刘广利. 不确定支持向量机原理及应用[M]. 北京: 科学出版社, 2007.

[47] 侯媛彬, 杜京义, 汪梅. 神经网络[M]. 西安: 西安电子科技大学出版社, 2007.

[48]董华,杨世元,吴德会. 基于模糊支持向量机的小批量生产质量智能预测方法[J]. 系统工程理论与实践,2007,3:98—102.

[49] Jayadeva, Reshma Khemchandani, Suresh Chandra. Fuzzy linear proximal support vector machines for multi-category data classification [J]. Neurocomputing,2005,67:426—435.

[50]H. P. Huang, Y. H. Liu. Fuzzy support vector machine for pattern recognition and data mining [J]. Int. J. Fuzzy Syst.,2002,4(3):826—835.

[51]C. F. Lin, S. D. Wang. Fuzzy support vector machines [J]. IEEE Transactions on Neural Networks,2002,13(2).

[52]Jeff Fortuna, David Capson. Improved support vector classification using PCA and ICA feature space modification [J]. Pattern Recognition,2004,37:1117—1129.

[53]杨世元,吴德会,苏海涛. 基于 PCA 和 SVM 的控制图失控模式智能识别方法[J]. 系统仿真学报,2006,18(5):1314—1318.

[54]王小平,曹立明. 遗传算法——理论、应用与软件实现[M]. 西安:西安交通大学出版社,2002.

[55]高尚,杨靖宇. 群智能算法及其应用[M]. 北京:中国水利水电出版社,2006.

[56]C. L. Huang, C. J. Wang. A GA-based feature selection and parameters optimization for support vector machines[J]. Expert Systems with Applications,2006,31:231—240.

[57]Shih-Wei Lin, Zne-Jung Lee, et al. Parameter determination of support vector machine and feature selection using simulated annealing approach [J]. Applied Soft Computing,2008,8:1505—1512.

[58]段海滨. 蚁群算法原理及其应用[M]. 北京:科学出版社,2005.

[59] J. Kennedy, R. Eberhart. Particle swarm optimization [A]. Proceeding IEEE International Conference onNneural Networks,1995,4:1942—1948.

[60] Y. Shi, R. Eberhart. A modified particle swarm optimizer [A]. IEEE World Congress on Computational Intelligence,1998,69—73.

[61] Yannis Marinakis, Magdalene Marinaki, Georgios Dounias. Particle swarm optimization for pap-smear diagnosis [J]. Expert Systems with Applications,2008,35:1645—1656.

[62] S. Hofmeyr, S. Forrest. Architecture for an artificial immune system [J]. Evolutionary Compution,1999,7(1):1289—1296.

[63] 李晓磊,邵之江,钱积新. 一种基于动物自治体的寻优模式:鱼群算法[J]. 系统工程理论与实践,2002,11:32—38.

[64] Cheng Chuen-Sheng, Norma Faris Hubele. Design of a knowledge-based expert system for statistical process control [J]. Computers and Industrial Engineering,1992,22(4):501—517.

[65] D. T. Pham, E. Oztemel. An on-line expert system for statistical process control [J]. International Journal of Production Research,1992,30(12):2857—2872.

[66] Min Soo Sub, Won Chul Jhee, et al. A case-based expert system approach for quality design [J]. Expert System With Applications,1998,15:181—190.

[67] F. T. S. Chan. In-line process conditions monitoring expert system for injection molding [J]. Journal of Materials Processing Technology,2000,101:268—274.

[68] 李刚,王霄,蔡兰. 基于专家系统的统计质量智能控制[J]. 江苏理工大学学报,2000,21(2):44—48.

[69] W. A. Shewhart. Economic control quality manufactured

products[M]. New York:Van Nostrand,1931.

[70]Western Electric. Statistical Quality Control Handbook (2d edition)[M]. New York:Mack Printing Company,1956.

[71] Abdelmonem Snoussia, Mohamed El Ghourabia, Mohamed Limama. On SPC for short run autocorrelated data [J]. Communications in Statistics-Simulation and Computation,2005,34(1):219－234.

[72] Maurice Pillet. A specific SPC chart for small-batch control [J]. Quality Engineering,1996,8(4):581－586.

[73]Mamoun Al-Salti,Anthony Statham. The application of group technology concept for implementing SPC in small batch manufacture [J]. International Journal of Quality & Reliability Management,1994,11(4):64－76.

[74]G. F. Koons, J. J. Luner. SPC in low volume manufacturing:a case study [J]. Journal of Quality Technology,1991,23(4):287－295.

[75]W. H. Woodall, B. M. Adams. The statistical design of CUSUM charts [J]. Quality Engineering,1993,5(4):559－570.

[76]A. F. Bissell. Cusum techniques for quality control [J]. Applied Statistics,1969,18:1－30.

[77]S. I. Chang,S. Y. Lin. A comparative study and Design of EWMA control charts for monitoring process variations in short run productions [J]. International Journal of Industrial Engineering,1996,3(4):268－278.

[78]袁哲俊,徐忡,马玉林. 面向 ATM 生产环境的 EWMA 质量控制图[J]. 哈尔滨工业大学学报,2000,32(1):45－50.

[79]张志雷. 自相关过程的改进型 EWMA 控制图[J]. 数理统计与管理,2008,27(3):466－472.

[80]孙静. 自相关过程的残差控制图[J]. 清华大学学报,2002,42(6):735－738.

[81]崔敬巍,谢里阳,刘晓霞.监视过程均值变化的残差控制图检测能力分析[J].东北大学学报,2007,28(3):401-404.

[82]张力健,杨继平,张秋菊.AR-GARCH型残差控制图及其应用研究[J].中国管理科学,2006,14:15-19.

[83]孙静,杨穆尔.多元自相关过程的残差T^2控制图[J].清华大学学报,2007,47(12):2184-2187.

[84] A. A. Kalgonda, S. R. Kulkarni. Multivariate quality control chart for autocorrelated process, Journal of Applied Statistics[J]. 2004,31(3):317-327.

[85]R. S. Guh. Robustness of the neural network based control chart pattern recognition system to non-normality [J]. International Journal of Quality & Reliability Management,2002,19(1):97-112.

[86] Wimalin Sukthomya, James Tannock. The training of neural networks to model manufacturing processes [J]. Journal of Intelligence Manufacturing,2005,16:39-51.

[87]AT&T. Statistical Quality Control Handbook (11th edition)[M]. Carolina:Delmar Printing Company,1985.

[88]D. T. Pham, M. A. Waini. Feature-based control chart pattern recognition [J]. International Journal Production Research,1997,35(7):1875-1890.

[89]乐清洪.智能工序质量控制的理论与方法研究[D].西北工业大学,2002.

[90]H. B. Hwarng, N. F. Hubele. Back-propagation pattern recognizers for X-bar control charts: methodology and performance[J]. Computer & Industrial Engineering,1993,24:219-35.

[91]R. S. Guh,J. D. T. Tannock. A neural network approach to characterize pattern parameters in process control charts [J]. Journal of Intelligent Manufacturing,1999,10:449-462.

[92]A. S. Anagun. A neural network applied to pattern rec-

ognition in statistical process control[J]. Computers & industrial engineering,1998,35(1-2):185-188.

[93]D. C. Reddy, K. Ghost. Identification and interpretation of manufacturing process patterns through neural networks [J]. Mathematical and Computer Modelling,1998,27(5):15-36.

[94]许东,吴铮. 基于 MATLAB 6.x 的系统分析与设计——神经网络[M]. 西安:西安电子科技大学出版社,2003.

[95]李国勇. 智能控制及其 MATLAB 实现[M]. 北京:电子工业出版社,2005.

[96]孙学静,刘飞. 基于 ART1 神经网络的统计过程控制系统 [J]. 控制理论与应用,2006,25(5):1-3.

[97]阎平凡,张长水. 人工神经网络与模拟进化计算[M]. 北京:清华大学出版社,2005.

[98]C. S. Cheng. A multi-layer neural network model for detecting changes in the process mean [J]. Computers & Industrial Engineering,1995,28(1):51-56.

[99]S. I. Chang, C. A. Aw. A neural fuzzy control chart for detecting and classifying process mean shift [J]. International Journal of Production Research,1996,34(8):2265-2278.

[100]M. B. Perry, J. K. Spoerre, T. Velasco. Control chart pattern recognition using back propagation artificial neural network [J]. International Journal of Production Research,,200139(15):3399-3418.

[101]R. S. Guh, F. Zorriassatine, et al. On-line control chart pattern detection and discrimination-a neural network approach [J]. Artificial Intelligence in Engineering,1999,13:413-425.

[102]吴德会,杨世元. 基于人工神经网络的 6σ 质量控制[J]. 航空制造技术,2007,3:83-86.

[103]R. S. Guh. A hybrid learning-based model for on-line detection and analysis of control chart patterns [J]. Computers &

Industrial Engineering,2005,49:35—62.

[104]乐清洪,滕霖,等.质量控制图在线智能诊断分析系统[J].计算机集成制造系统,2004,10(12):1583—1587.

[105]D. F. Specht. Probabilistic neural networks for classification,mapping,or associative memory [C]. IEEE International Conferenceon Neural Networks,1988,1:525—532.

[106]D. F. Specht. Probabilistic neural network [J]. Neural Network,1990,3(2):109—118.

[107] B. Ertholdm, D. Iamondj. Constructive training of probabilistic neural networks [J]. Neurocomputing,1998,19(1—3),167—183.

[108]边肇祺,张学工.模式识别[M].北京:清华大学出版社,2000.

[109]Sergios Theodoridis,Konstantinos Koutroumbas.模式识别[M].北京:电子工业出版社,2006.

[110]P. D. Wasserman. Advanced Methods in Neural Computing [M]. New York:John Wiley & Sons,Inc. ,1993.

[111]D. F. Specht. Enhancements to probabilistic neural networks [C]. International Joint Conference on Neural Networks,1992,1:761—768.

[112]克莱斯勒汽车公司,福特汽车公司,通用汽车公司.统计过程控制[M].上海:美国汽车工业行动集团,2005.

[113]A. A. Yousef. Recognition of control chart patterns using multi-resolution wavelets analysis and neural networks [J]. Computers and Industrial Engineering,2004,47:17—29.

[114]吴少雄.智能统计工序质量控制的体系研究[J].计算机集成制造系统,2006,12(11):1832—1837.

[115]Verado Carmo C. de Vargas,Luis Felipe Dias Lopes, Adriano Mendonc a Souza. Comparative study of the performance of the CuSum and EWMA control charts [J]. Computers & In-

dustrial Engineering,2004,46:707－724.

[116]殷建军,项祖丰,叶力.多元混合分布的EWMA控制图的平均链长[J].浙江工业大学学报,2006,34(1):65－68.

[117]余忠华,吴昭同.面向多品种、小批量制造环境的SPC实施模型的研究[J].工程设计,2000,4:28－31.

[118]孙群,赵颖,孟晓风.基于小样本数据特征的测量过程控制图参数优化[J].系统仿真学报,2008,20(11):2987－2990.

[119]Andrew R Webb. Statistical Pattern Recognition [M]. New York:John Wiley & Sons Ltd,2002.

[120]J. A. K. Suykens. Nonlinear Modelling and Support Vector Machines [C]. IEEE Instrumentation and Measurement Technology Conference,Budapest,Hungary,2001,pp. 287－294.

[121]C. Cortes, V. Vapnik. Support vector networks [J]. Machine Learning,1995,20:273－297.

[122]N. Cristianini,J. T. Shawe. An Introduction to Support Vector Machines [M]. London:Cambridge University Press,2000.

[123] V. David Sanchez A. Advanced support vector machines and kernel methods [J]. Neurocomputing,2003,55,5－20.

[124]Suykens Johan A K,Gestel Tony Van,Brabanter Jos De. Least squares support vector machines [M]. Singapore:World Scientific Publishers,2002.

[125]J. A. K Suykens,J. Vandewalle. Least squares support vector machine classifiers [J]. Neural Processing Letters,1999,9,293－300.

[126]陈爱军,宋执环,李平.基于矢量基学习的最小二乘支持向量机建模[J].控制理论与应用,2007,24(1),1－5.

[127]宋海鹰,桂卫华,阳春华.稀疏最小二乘支持向量机及其应用研究[J].信息与控制,2008,37(3),334－338.

[128]G. C. Cawley, N. L. C. Talbot. Improved sparse least-squares support vector machines [J]. Neurocomputing Letters, 2002, 48:1025－1031.

[129]F. R. Hampel, E. M. Ronchetti, P. J. Rousseeuw, W. A. Stahel, Robust Statistics: The Approach Based on Influence Functions[M]. New York: Wiley, 1986.

[130]C. W. Hsu, C. J. Lin. A comparison of methods for multi-class support vector machines[J]. IEEE Trans Neural Networks, 2002, 13:415－425.

[131]E. J. Bredensteiner, K. P. Bennett. Multicategory classification by support vector machines [J]. Comput. Optimiz. Applicat., 1999, pp. 53－79.

[132]U. H. G. Krebel. Pairwise classification and support vector machines [J]. In: Scholkopf B, Burges C, Somla A (eds) Advances in kernel methods: support vector machine. MIT Press, Cambridge, MA, 1998, pp. 255－268.

[133]Bo Liu, Zhifeng Hao, Xiaowei Yang. Nesting algorithm for multi-classification problems [J]. Soft Computing, 2007, 11: 383－389.

[134]V. G. Tony, et al. Benchmarking least squares support vector machine classifiers [J]. Machine Learning, 2004, 54(1): 5－32.

[135]Senjian An, Wanquan Liu, et al. Fast cross-validation algorithms for least squares support vector machine and kernel ridge regression [J]. Pattern Recognition, 2007, 40(8):2154－62.

[136]G. C. Cawley, N. L. C. Talbot. Fast exact leave-one-out cross-validation of sparse least-squares support vector machines [J]. Neural Networks, 2004, 17(10):1467－1475.

[137]郭辉,刘贺平,王玲. 最小二乘支持向量机参数选择方

法及其应用研究[J]. 系统仿真学报,2006,18(7):2033－2036.

[138] W. E. Deming. Out of Crisis[M]. Cambridge,MA: MIT Press,1986.

[139]ZHAO NING HAN DONG. Comparison of CUSUM, GLR,GEWMA and RFCuscore in Detecting Mean Shifts of Stable Processes [J]. Chinese Journal of Applied Probability and Statistics,2005,21(4):403－411.

[140] H. B. Nembhard, R. V. Ventura. Cuscore statistic to monitor a non-stationary system [J]. Quality and Reliability Engineering International,2007,23:303－325.

[141]R. A. Fisher. Theory of statistical estimation [A]. Proceedings of the Cambridge Philosophical Society,1925,22:700－725.

[142]G. E. P. Box,J. Ramirez. Cumulative score charts [J]. Quality and Reliability Engineering International,1992,8:17－27.

[143] H. B. Nembhard, Shuohui Chen. Cuscore Control Charts for Generalized Feedback-control systems [J]. Quality and Reliability Engineering International,2007,23:483－502.

[144] P. Changpetch, H. B. Nembhard. Periodic Cuscore Charts to Detect Step Shifts in Autocorrelated Processes [J]. Quality and Reliability Engineering International,2008,24:911－926.

[145] H. B. Nembhard. Cuscore statistics:Directed process monitoring for early problem detection[M]. Handbook of Engineering Statistics,Pham H (ed.). Springer:Berlin,2006.

[146]G. E. P. Box,Luceño A. Statistical Control by Monitoring and Feedback Adjustment [M]. Wiley:New York,1997.

[147] L. Shu, D. W. Apley, F. T. Sung. Autocorrelated Process Monitoring Using Triggered Cuscore Charts [J]. Quality

and Reliability Engineering International,2002,18:411—421.

[148] H. B. Nembhard. Simulation using the state-space representation of noisy dynamic systems to determine effective integrated process control designs [J]. IIE Transactions,1998,30: 247—256.

[149] D. M. Hawkins, P. Qiu, C. W. Kang. The changepoint model for statistical process control [J]. Journal of Quality Technology,2003,35:355—365.

[150] K. D. Zamba, D. M. Hawkins. A multivariate change point model for statistical process control [J]. Technometrics, 2006,48(4):539—549.

[151] K. D. Zamba. A multivariate change-point model for change in mean vector and/or Covariance Structure [J]. Journal of Quality Technology,2009,41(3):285—303.

[152] Y. Ninomiya. Information criterion for Gaussian change-point model [J]. Statist. Probab. Lett. ,2005,72:237—247.

[153] Jianmin Pan, Jiahua Chen. Application of modified information criterion to multiple change point problems [J]. Journal of Multivariate Analysis,2006,97:2221—2241.

[154] Gregory Gurevich, Albert Vexler. Change point problems in the model of logistic regression [J]. Journal of Statistical Planning and Inference,2005,131:313—331.

[155] George Box, Tim Kramer. Statistical process monitoring and feedback adjustment-a discussion [J]. Technometrics, 1992,34(3):251—267.

[156] F. Tsung, K. L. A. Tsui. A mean shift pattern study on integration of SPC and APC for process monitoring [J]. IIE Transactions,2003,35:231—242.

[157] A. Luceno. Cuscore charts to detect level shifts in au-

tocorelated noise [J]. Quality Technology and Quantitative Management, 2004, 1(1): 27－45.

[158] Y. E. Shao. Integrated application of the cumulative score control chart and engineering process control [J]. Statistica Sinica, 1998, 8: 239－252.

[159] Herriet Black Nembhard, Pannapa Changpetch. Directed monitoring using cuscore charts for seasonal times series [J]. Quality and Reliability Engineering International, 2007, 23: 219－232.

[160] J. Ramirez. Monitoring clean room air using cuscore charts [J]. Quality and Reliability Engineering International, 1992, 14: 281－289.

[161] H. Hotelling. Multivariate quality control [J]. Techniques of Statistical Analysis, 1947, 113－183.

[162] 张杰, 阳宪惠. 多变量统计过程控制 [M]. 北京: 化学工业出版社, 2000.

[163] Theodora Kourti. Application of latent variable methods to process control and multivariate statistical process control in industry [J]. International Journal of Adaptive Control and Signal Processing, 2005, 19: 213－246.

[164] L. H. Chiang, L. F. Colegrove. Industrial implementation of on-line multivariate quality control [J]. Chemometrics and Intelligent Laboratory Systems, 2007, 88: 143－153.

[165] H. Albazzaz, X. Z. Wang. Statistical process control charts for batch operations based on independent component analysis [J]. Industrial & Engineering Chemistry Research, 2004, 43 (21): 6731－6741.

[166] Manabu Kano, Shinji Hasebe, et al. Evolution of multivariate statistical process control: application of independent component analysis and external analysis [J]. Computers and Chemi-

cal Engineering,2004,28:1157—1166.

[167] F. Aparisi, C. L. Haro. Hotelling's T^2 control chart with variable sampling intervals [J]. International Journal of Production Research,2001,39:3127—3140.

[168] F. B. Alt. Multivariate quality control. Encyclopedia of the Statistical Science, Johnson NL, Kotz S, Read CR (eds.) [M]. Wiley:New York,1985,6:111—122.

[169] J. E. Jackson. Multivariate quality control [J]. Communications in Statistics—Theory and methods,1985,14:2657—2688.

[170] R. B. Crosier. Multivariate generalizations of cumulative sum quality control schemes [J]. Technometrics,1988,30:291—303.

[171] C. A. Lowry, D. C. Montgomery. A review of multivariate control charts [J]. IIE Transactions,1995,27:800—810.

[172] C. A. Lowry, W. H. Woodall, C. W. Champ, S. E. Rigdon. A multivariate exponentially weighted moving average control chart [J]. Technometrics,1992,34:46—53.

[173] R. L. Mason, C. W. Champ, N. D. Tracy, S. J. Wierda, J. C. Young. Assessment of multivariate process control techniques [J]. Journal of Quality Technology,1997,29:140—143.

[174] W. H. Woodall. Control charts based on attribute data:Bibliography and review [J]. Journal of quality Technology,1997,29(2):172—183.

[175] W. H. Woodall, M. M. Ncube. Multivariate CUSUM quality control procedures [J]. Technometrics,27:285—292,1985.

[176] M. A. G. Machadol, A. F. B. Costal, M. A. Rahim. The synthetic control chart based on two sample variances for monitoring the covariance matrix [J]. Uality and Reliability Engineering

International,2008,25(5):595—605.

[177]张公绪. 两种质量诊断理论及其应用[M]. 北京:科学出版社,2001.

[178]J. E. Jackson. User Guide to Principal Components[M]. Wiley:New York,1991.

[179]A. J. Hayter, K. L. Tsui. Identification and quantification in multivariate quality control problems [J]. Journal of Quality Technology,1994,26:197—208.

[180]Hassen Taleb. Control charts applications for multivariate attribute processes [J]. Computers & Industrial Engineering,2009,56:399—410.

[181]Zhou Chunguang, Zou Changliang, Wang Zhaojun. A robust control chart based on wavelets for preliminary analysis of individual observations [J]. Chinese Journal of Applied Probability and Statistics,2008,24(3):274—288.

[182]R. Noorossana, M. Farrokhi, A. Saghaei. Using neural networks to detect and classify out-of-control signals in autocorrelated processes [J]. Quality and Reliability Engineering International,2003,19:493—504.

[183]Seyed Taghi Akhavan Niaki, Babak Abbasi. Fault diagnosis in multivariate control charts using artificial neural networks [J]. Quality and Reliability Engineering International,2005,21:825—840.

[184]Long-Hui Chen, Tai-Yue Wang. Artificial neural networks to classify mean shifts from multivariate χ^2 chart signals [J]. Computers & Industrial Engineering,2004,47:195—205.

[185]R. S. Guh. On-line identification and quantification of mean shifts in bivariate processes using a neural network-based approach [J]. Quality and Reliability Engineering International,2007,23:367—385.

[186] Jian-bo Yu, Li-feng Xi. A neural network ensemble-based model for on-line monitoring and diagnosis of out-of-control signals in multivariate manufacturing processes [J]. Expert Systems with Applications,2009,36:909—921.

[187] Chinyao Low, Chih-Ming Hsu, Fong-Jung Yu. Analysis of variations in a multi-variate process using neural networks [J]. International Journal of Advanced Manufacture Technology, 2003,22:911—921.

[188] Ben Khediri Issam, Limam Mohamed. Support vector regression based residual MCUSUM control chart for autocorrelated process [J]. Applied Mathematics and Computation,2008, 201:565—574.

[189] Chuen-Sheng Cheng, Hui-Ping Cheng. Identifying the source of variance shifts in the multivariate process using neural networks and support vector machines [J]. Expert Systems with Applications, Elsevier Ltd. In Press,2007.

[190] D. C. Montgomery. Introduction to Statistical Quality Control [M]. New York:John Wiley & Sons,1991.

[191] J. E. Jackson. A User's Guide to Principle Components[M]. New York:John Wiley & Sons,1991.

[192] S. Bersimis, S. Psarakis, J. Panaretos. Multivariate statistical process control charts:an overview [J]. Quality and Reliability Engineering International,2007,23:517—543.

[193] F. Zorriassatine, J. D. T. Tannock, C. O. Brien. Using novelty detection to identify abnormalities caused by mean shifts in bivariate processes [J]. Computers & Industrial Engineering, 2003,44:385—408.

[194] D. C. Montgomery. Introduction to Statistical Quality Control[M]. New York:Wiley,2004.

后 记

本专著是作者博士研究工作的总结与拓展,除了作者本人的努力之外,还得到了其他老师、同事和同学的无私帮助与指导。本书的完成首先要感谢我的导师马义中教授。马老师治学严谨、学识渊博、求真务实、勇于创新,对学术研究和教学工作兢兢业业,一丝不苟。学术研究上,马老师从研究方向与选题、研究方法等方面给予我悉心的指导和帮助,并尽其所能创造和提供良好宽松的学术学习环境与科研条件;生活上,马老师给予关心照顾,体谅学生的困难并给予帮助;组织上,关心学生的进步与前途。尤其是马老师宽宏待人、谦虚儒雅的做人做事风格,深深影响并感动着我,对我以后的人生道路产生教育和激励的作用。在此本书完成之际,我由衷感谢马老师对我的培养和关怀,感谢他为我的成长付出的诸多心血!

还要感谢南京理工大学经济管理学院的冯俊文老师、薛恒新老师、杨文胜老师、陈杰老师、宋华明老师以及南京理工大学自动化学院的孙金生老师、孔建寿老师,加拿大卡尔加里大学的涂忆柳老师等人在我的研究思路和方法上的指导,他们的指导和点拨大大地提高了我对学术问题的研究与认识水平。

要感谢质量工程课题组的刘利平博士、邓海松博士、汪建均博士、周晓剑博士以及刘宝翠硕士、刘斌硕士、刘阳硕士、郭颖硕士、田甜硕士、苏国进硕士等人,感谢南京理工大学经济管理学院同年级的苗成林博士、郭志芳博士、谢延浩博士、陈成博士、柏菊博士等人,读博士四年以来,他们为我多处排忧解难。感谢我的室友牛军博士、卜京博士、徐俊博士、万明华博士、李羊城博士,以及其他学院的任贵州博士、邵杰博士、黄传波博士、叶树霞博士等人,他们的友谊和帮助使我在其他学科的知识和方法中受益匪浅,并使我度过了虽然艰苦却是非常有意义的四年博士学习和研究生涯。

感谢我妻子和儿女对我攻读博士学位的理解和支持,感谢我的亲朋好友对我读博士期间给予我个人及家庭的各种各样的帮助。还要感谢我的父母双亲,他们真诚和朴实的一生,使我懂得了勤奋与坚韧的力量。

作 者

2017年10月